Urban Sustainability

Editor-in-Chief

Ali Cheshmehzangi ⓘ, Architecture & Built Environment, University of
Nottingham Ningbo China, Ningbo, Zhejiang, China

The Urban Sustainability Book Series is a valuable resource for sustainability and urban-related education and research. It offers an inter-disciplinary platform covering all four areas of practice, policy, education, research, and their nexus. The publications in this series are related to critical areas of sustainability, urban studies, planning, and urban geography.

This book series aims to put together cutting-edge research findings linked to the overarching field of urban sustainability. The scope and nature of the topic are broad and interdisciplinary and bring together various associated disciplines from sustainable development, environmental sciences, urbanism, etc. With many advanced research findings in the field, there is a need to put together various discussions and contributions on specific sustainability fields, covering a good range of topics on sustainable development, sustainable urbanism, and urban sustainability. Despite the broad range of issues, we note the importance of practical and policy-oriented directions, extending the literature and directions and pathways towards achieving urban sustainability.

The series will appeal to urbanists, geographers, planners, engineers, architects, governmental authorities, policymakers, researchers of all levels, and to all of those interested in a wide-ranging overview of urban sustainability and its associated fields. The series includes monographs and edited volumes, covering a range of topics under the urban sustainability topic, which can also be used for teaching materials.

Qinran Yang

Gentrification in Chinese Cities

State Institutions, Space and Society

 Springer

Qinran Yang
Department of Urban and Rural Planning,
School of Architecture
Southwest Jiaotong University
Chengdu, China

The Youth Program of National Natural Science Foundation of China (51808452), The Sichuan Planning Program of Philosophy and Social Sciences (SC20B113) and The Doctoral Scholarship of the UBC-China Scholarship Council (201206050015)

ISSN 2731-6483   ISSN 2731-6491 (electronic)
Urban Sustainability
ISBN 978-981-19-2288-6   ISBN 978-981-19-2286-2 (eBook)
https://doi.org/10.1007/978-981-19-2286-2

This Springer imprint is published by the registered company Springer Nature Singapore Pte Ltd.
The registered company address is: 152 Beach Road, #21-01/04 Gateway East, Singapore 189721, Singapore

Preface

Gentrification continues to be a central issue in urban studies for more than a half century. The phenomenon becomes a beneficial perspective to scrutinize the comparative modes of urbanization among metropolises in different regions. For gentrification is often set off by forces of globalization, but could be rewritten by regional contexts and local settings in its development paths, manifestations, effects and even meanings. The global and provincial attributes of gentrification determine its indeterminacy. It appears in multiple forms and attracts a variety of storytelling about it and its mutations.

The politics of China is arguably one of the central conditions that mystifies the road of Chinese urbanization. State institutions relating to socio-spatial transformation in a transitional economy further complicate the trajectory of urban change. This book takes institutional change as a guiding line to examine the intricate processes shaping gentrification in Chinese cities, mainly during the 2000s. It faces directly the processual, multifaceted and path-dependent characters of gentrification and treats institutional making as an intermediary that, on the one hand, responds to the global trends and, on the other hand, adapts to local preconditions. The book divides gentrification into its constituent parts of socio-spatial changes that have been promoted by the changing cultural, economic and social institutions in spatial production. Chengdu, an emerging global city, is selected as a primary case study. The municipal government's mobilization in developmental strategies concentrating on the built environment allows the case study to make new claims for gentrification knowledge for non-Western cities. Presenting an institutional interpretation of gentrification in the case city, the book is hopefully to advance its understanding in China and the comparative studies among the Western and non-Western cities.

The conception of gentrification concerns about essentially the patterns of transformation or reproduction of urban social spaces. The relationship between state, space and society is thus a central axis of the institutional interpretation of gentrification in this book. It provides a general review of post-socialist societal transition in Chinese large cities, upon which the institutional changes navigating the proceeding of gentrification are scrutinized from an interactional approach between the state and social actors involved in the processes. How do the state actors interact with the middle-class consumers, directing the socio-spatial upgrading in the Chinese cities? How do the state actors interact with the lower-income social groups, shifting their housing opportunities in the cities? How do these interactions create a new politics of gentrification, that is, the good and evil of gentrification? The book will enrich an understanding of the Chinese distinctions of gentrification from a socio-political perspective. I bear in mind that the state–society relations occupy a considerable, rooted and elusive part of the Chinese characteristics, which is a necessary start to the storytelling of gentrification in China.

Chengdu, China   Qinran Yang

The original version of the book frontmatter was revised: A wrong word added in the Funding Information text has been updated. The correction to the book is available at https://doi.org/10.1007/978-981-19-2286-2_11

Acknowledgements

I acknowledge the National Natural Science Foundation of China for funding the project "Influencing Mechanism of Spatial Pattern of Resettlement Areas on the Reconstruction of Neighborhood Social Network of Relocated Residents: The Example of Caojiaxiang Project in Chengdu" (51808452), the Sichuan Federation of Social Science Associations for funding the project "Research on the Urban Development Strategy and Gentrification in Chengdu against the Background of Inland Opening" (SC20B113) and the UBC-China Scholarship Council for funding my doctoral program.

I am deeply indebted to my doctoral supervisor, Professor David Ley, who has been supportive of the research and worked actively to help me in pursuing the research goals. I have benefited from his insights into gentrification and the way he builds knowledge for Asian cities, which guided me to understand and develop ideas for the urban and social change in China. The work would not have been possible without his assistance in the writing process. His critical reading and guidance in the refinement of ideas have enabled me to achieve higher quality. I cannot express enough thanks to Prof. Chris Hamnnet, Prof. Ali Cheshmehzangi, Prof. Xing Huang and Prof. Weixuan Song for reviewing the entire manuscript and for the excellent advice. I offer my sincere appreciation for the learning opportunities provided by them.

The completion of this project owes a good deal to a large number of interviewees, both urban dwellers and specialists from Chengdu. I thank my informants for sharing experiences and opinions with me. I have enjoyed and learnt from our conversations. I am also grateful to the community and institutional members who introduced me to the interviewees and aided me in collecting materials during my fieldwork in Chengdu.

My debt of gratitude is to my families and friends who keep me in the pursuit of an academic career. My parents, who are always proud of me, made me brave when the times got tough. Benying Xu, the partner in my life, has protected my dream with his love.

Contents

About the Author

Qinran Yang is currently a lecturer in the Department of Urban and Rural Planning in Southwest Jiaotong University, China. She received bachelor and master's degrees in urban planning from Chongqing University and a Ph.D. degree in human geography from the University of British Columbia. She is interested in urban regeneration, gentrification and housing and community planning. She is recently working on two research projects, *"Urban Development Strategy and Gentrification in Chengdu against the Background of Inland Opening"* and *"Influencing Mechanism of Spatial Pattern of Resettlement Areas on the Reconstruction of Neighborhood Social Network of Relocated Residents"*. More projects included the *"Urban governance in neighborhood renewal"* and *"The politics of state-facilitated gentrification in post-socialist China"*.

List of Figures

List of Tables

Chapter 1
Introduction

It was the winter of 2014. A middle-aged man was standing in front of a redbrick building that was being demolished. He said, in a self-mocking tone, "Here was once the fanciest residence in the city. In the 1950s, we were the largest state-owned construction company in the city. We founded the first brickyard in the city and used the bricks they produced to build those redbrick buildings. We invited Soviet designers to design the architecture. Now we have abandoned them. Evergrande (the developer) will become its master. Those *danwei* (work-unit) people will finally move out of the housing. Perhaps in the future, people living here may not know that this was once the property of Huaxi Group (the name of the work-unit)." The middle-aged man is a former manager of a subsidiary real estate corporation of Huaxi Group. However, he was appointed to join a state-owned financing platform company that is directly affiliated with the district government and is charged with housing demolition and compensation. He spoke those words in response to being asked why Huaxi Group, as the actual land user, has not sought an alternative for renovating the neighborhood by which it may retain its land-use rights. His answers show that, inevitably, a new regime of urban construction is now emerging.

The landscape of declining old neighborhoods and emerging spectacular images in inner cities exhibits a dramatic shift in the power structure of agents on inner-city terrains. The disappearance of declining old neighborhoods signifies the abolishment of the socialist urban regime on land development and spatial production, while the emerging spectacular images mark the introduction of a new urban regime that features a so-called market economy with Chinese characteristics. The housing system affiliated with socialist *danwei* has finally ended, together with the residual low-income *danwei* workers that have, either voluntarily or otherwise, been incorporated into the new housing system.

The landscape displays the normative gentrification process occurring in large Chinese cities especially during the 2000s, which portrays the shift in class power from the lower strata currently located in a place to a higher social strata. However, the scene also discloses an obvious but less detectable characteristic of gentrification in Chinese cities—changing social power over places is sourced from and conditioned

© Springer Nature Singapore Pte Ltd. 2022
Q. Yang, *Gentrification in Chinese Cities*, Urban Sustainability,
https://doi.org/10.1007/978-981-19-2286-2_1

by institutional transition in post-reform spatial production, in its cultural, economic and social aspects.

This characteristic is evident from the unique nature of old neighborhoods in Chinese cities. Neighborhood renewal in China mainly involves three types of old neighborhoods/housing in cities: neighborhoods/housing for work-unit residents, municipal neighborhoods/housing for nonwork-unit residents, and urban villages. Rather than mere living terrain or potential supply/demand markets, these neighborhoods are essentially institutional products of a socialist redistributive economy, social organization and socialist modernity. In this case, neighborhood renewal has not simply denoted land reinvestment but has also been related to the new institutional establishment of a new land market and housing provision system in China. It has not merely transformed the pattern of social life but also led to modern construction in post-socialist China. Most importantly, neighborhood renewal has not straightforwardly resulted in residential change but has subtly reinvented the meanings of properties to various social groups. Thus, the promotion of large-scale neighborhood renewal per se has formed part of the "path" toward institutional transition and social change that has unfolded in urban China.

The primary contribution of this study lies in a systematic analysis of how the specific dynamics of institutional change have altered the power of various social groups over places. It questions the sets of ideas and rules that initiate and guide gentrification in Chinese cities. How are the ideas and rules made? How do these ideas and rules determine the social and spatial results of gentrification with certain regional characteristics? Moreover, as the core connotation of gentrification has been social power shift in place, this study intends to approach the roles of the state and society equally in the institutional change in gentrification. A central assumption throughout the book is that the relations between the state and society, rather than either the state or the society acting in isolation, shape the characteristics of the gentrification process in China. This analysis is skeptical both of the structural explanation and of the revelation of unsorted differences; instead, it aims to generate midrange regularities of gentrification in Chinese cities that cohere with context.

1.1   The Global Versus Local Debate

In 2002, Smith argued that gentrification was now worldwide, spreading from the Global North to the Global South and showing remarkable growth in metropolitan areas. Economic globalization and neoliberal state restructuring are the new forces driving gentrification on a global scale. The prevalence of neoliberal political ideology is also creating a common global trend of gentrification in relation to the role of the state. Hackworth and Smith defined this phenomenon as a third wave of gentrification that began after the 1987 recession in European and North American cities; this wave involves far more aggressive action on the part of real estate industries and greater state stimulation of the free market [45, 120]. These neoliberal urban strategies lure the middle class back to the city [72, 119] and strengthen the tax base and labor market potential, they absorb the entry of high-end service industries, they increase foreign direct investment in industrial and real estate development, and they stimulate city consumption through urban spectacles [50, 51]. In particular, by creating a promising business climate and comprehensive urban facilities, urban regeneration is likely to upgrade the bond rating of the metropolis as assessed by international bond-rating agencies. Thus, the city holds potential advantages in its ability to attract more investors in government bonds for public service delivery [44, 72].

From a social and cultural perspective, another set of global postulates is arising in the context of cosmopolitanism and transnational urbanism. The cosmopolitan population and culture promote an urban imagery that is harnessed by those local ruling elites who make decisions in favor of gentrification; meanwhile, gentrification is now based on so-called neoliberal ideas. In this global context, the connection between cultural ideology, urbanism and middle-class mobility implied in gentrification could be reinterpreted to indicate that the state, transnational capitalists and a new social group composed of a cosmopolitan population have been the main agents initiating the process. Mega-projects are launched by the state, which aims to deliver a global image for its cities to attract expatriates and domestic professionals and managers [6, 111]. Even in metropolitan regions possessing limited competitive advantages in the global economy, governments at the national and subnational levels could, perhaps consciously, take as an imperative the need to "sell" a world-city landscape to transnational investors and consumers dreaming of a cosmopolitan lifestyle. Accordingly, the cosmopolitan population also reshapes the neighborhood landscape and socioculture to establish a sense of place [14, 15] and reconstruct class identities [16, 69].

Lees et al. [71] proposed a relational comparative approach to global gentrification. Seeking global regularities, these authors asserted that gentrification has become globalized alongside capital globalization, defined as capital shifting from industries to the secondary circuit of accumulation "in forms of fixed capital and, more specifically, the built environment and urban space" ([71], p. 45). Class struggle in gentrification, in this case, is treated as an abiding feature of a capitalist society. Lees et al. [71] proposed a thesis for the new economics of gentrification to describe the process of reinvestment, exploitation and accumulation. This thesis advanced Smith's rent-gap theory to explain existing divergences in the land reinvestment process in the Global South. Based on the Southern experience, for instance, the authors found that multiple actors could capture capitalized ground rent and participate in a share of the rent gap generated by redevelopment. Additionally, economic actors, state and authoritative instruments could impact how rent gaps were created through practices based on state-led, discursive practices, such as place stigmatization. The global theory advanced here develops Smith's structural explanation, in which gentrification is invariably induced by the cycle of uneven development intrinsic to capitalism and promoted by champions of capital accumulation through rent-gap exploitation. These global regularities have been solely based on the presumed situation of global capitalism in real estate industries, while social and cultural agencies, locally and internationally, are secondary issues.

The above global theories of gentrification have been challenged by scholars who treat the variation of the Global South experience of vital importance. These authors have uncovered a variety of contextualized differences between regions, requiring new approaches and explanations. The differences have been approached through, first, the structural forces and dominant power in promoting regional gentrification. The role of the state, for example, is commonly used by scholars to justify their alternative interpretations (see [40, 49]). Shin et al. [115] argued that, generally, in the Global East, states have been associated with capitalism and have engaged in real estate development as a means of urbanization, industrialization and state-building. In contrast, booming middle-class suburbanization, urban policies that favor the suburban real estate market, and the so-called "neglectful" behavior of the state in governing the urban core and making policy concessions for revitalization have limited the efficacy of gentrification [34, 150]. Second, a major stream within the current literature treats the flexible methods behind land and housing transformation in cities in the Global South as the starting point for questioning classical explanations. These include, for example, the bottom-up process of housing marketization in state-subsidized settlements in South Africa [74]; state planning and intervention in land marketing and housing provision [7, 150]; rent regulation by local governments and the prevention of gentrification [66, 123], and land enclosure, privatization and formalization in the less-established private property system in India [32, 41].

Third, a relatively small body of work has focused on social and cultural actors in less advanced economies. Some of these studies have treated these actors as an intricate factor likely to impede gentrification, as the new middle class has not reached a volume equal to that in the metropolitan cities in developed postindustrial societies [9]. Other studies examine the sophisticated sociocultural factors in localities that can make the gentrification process less than decisive in urban transitions, such as unique property conditions and residential patterns [90, 91], diverse tenure types and different communities of interest [41]. Fourth, the existing literature questions the ambiguity of the social outcomes of gentrification among the affected residents, and has uncovered both a substantial scale of class turnover [63, 67] and a neutral attitude toward urban redevelopment (i.e., the gains or losses of different residents in similar processes [76, 137]). As a result, the gentrification scholarship has become increasingly fragmented, and a large number of modifications of the classical gentrification modalities are emerging, such as studentification and rural, tourist, green and hybrid gentrification.

Among these studies, the literature on China has exposed three central features of gentrification in Chinese cities. First, the Chinese central and local states play important roles throughout the process of spatial (re)production and residential relocation [48, 52, 59, 115, 142]. Second, and related, thorough economic and cultural institutional transitions codify the processes of redevelopment and residential relocation in the inner city, led by the new market-oriented urban regime [54, 97, 112, 155]. Last, the redevelopment and residential relocation led by politico-economic agents should accommodate the intricate social conditions and dynamics. Substantial socioeconomic and sociocultural transformation determine the attributes of the middle class in China, which renders the sources and motives of gentrifiers more or less indistinct [53, 105, 133, 154]. Nevertheless, impacted by the reform strategies of housing and other factors, decision-making on residential relocation can produce variable outcomes among the diverse social groupings of the affected residents [33, 62, 82, 113, 114, 141].

Therefore, while structuralist scholars are committed to explaining the commonalities among the driving forces of gentrification, localist scholars are pushing gentrification toward fragmentation. The structuralist interpretations reveal the common characteristics of the major metropolises in global city-making, market-oriented systemic transformation, and transnational capitalism. However, from both politico-economic and sociocultural perspectives, the analytical frameworks fall short of capturing the entire course of gentrification. The fragmented process of theoretical building has reduced its implications for local urban change. Empirical studies on the Global South have enriched the trajectories of urban and social change, but they have also shattered any illusion of homogeneity and incapacitated a global theory that embraces regional differences. These studies have also increased the elusiveness of the concept by overloading it with untraceable variations.

This book suggests an institutional explanation for gentrification to generate midrange regularities that cohere to the context. According to the three schools of new institutionalism (i.e., rational choice, sociological and historical institutionalism), institutions include both formal rules and informal norms (e.g., values, customs and conventions). Institutional structure defines the way of human interactions and actions; moreover, institutions provide incentives to players [46, 98]. Institutions in this book refer to the ideas, norms, and rules that regularize, formally and informally, activities around spatial production and residential relocation. These include not only changing economic regulations but also cultural ideas in place-making and social policies for the residents affected by gentrification. An institutional analysis offers an important lens for a historical review of the trajectory of urban change. This method helps researchers probe the complexities of urban processes in a transitional economy. As Steinmo [121] has clarified, historical institutionalism sits at the mesolevel of the approaches to social science. A mesoapproach such as historical institutionalism does not assume that real-world outcomes can be functionally explicable by the overarching structure because contexts exert influences on decisions and outcomes. An institutional analysis thus does not stress the who and why of urban redevelopment so much as how projects unfold in practice. The analysis is instrumental to disentangling the interactions among actors during the process of urban redevelopment, with an emphasis on the path of state intervention in economic, cultural and societal spheres.

Historical institutionalists define "path dependence" and "unintended consequences" as basic characteristics of institutional change ([46], p. 938). In the processes of path-dependent institutional change, the forces and agents that fuel institutional change are mediated by given contextual conditions that have been generated by past institutional arrangements. The path-dependent feature of institutional change necessitates an interactional approach to the state, market and social players leading the course of change. As the current work has clearly expanded on the often collusive relationship between the state and capitalists in the process of land marketization [52, 55, 59, 111, 112, 160], it has rendered passive the role of sociocultural agents in gentrification. This study elaborates on the social involvement of gentrifiers and lower-income residents in gentrification.

The next two sections help to establish the explanatory frame of an institutional interpretation of state-facilitated gentrification in Chinese cities. The cultural, economic and social aspects of institutional change that give substance to the ideas and practices of the central and local governments are identified and associated with the social dynamics of a gentrification process. Mainly, in the process of socio-spatial upgrading, the cultural ideas of the local state in urban construction and the commodification of inner-city spaces are linked to the subject construction of middle-class consumer citizens. In the displacement process, the policy-making of rehousing the low-income residents and the governance of social confrontation are related to the social re-stratification of affected residents in gentrification.

1.2   Cultural Urbanism, Spatial Commodification and Consumer Class-Making

State-led urban redevelopment in China serves a wider objective than the immediate purpose of land income, such as industrial transformation, tourist development, consumption stimulation and spatial formalization [142]. Against these diverse objectives, the state-facilitated creation of new cultural urbanism and spatial commodification have directly determined the spatial production of gentrification. These two aspects of spatial practices are then associated with the consumption of gentrified places by middle-class consumers. This section connects three bodies of literature to explain the socio-spatial upgrading process fulfilled together by political economic and sociocultural actors in Chinese cities.

1.2.1   City Branding

Gentrification in China occurs in the context of the transformation from the modernity of the production society in socialist China to that featuring a rising consumer society. Ong [99] explains modernity as "an evolving process of (social) imagination and practice in particular historically situated formations"; "the social imaginaries have been called the 'constructed landscapes of collective aspirations' ([3], p. 5)" (p. 171). In regard to the city, Davis [28] terms Mao's vision of modernity as a *decommodified modernity* defined by "increased industrial output and the triumph of collective ownership" (p. 698). Cartier [19–21], in contrast, deems the landscape formation of transitional urbanism to be in line with the creation of consumer modernity and the promotion of the consumer economy by the state. In this context, state-led urban redevelopment is a spatial practice embodying ideologies and urban imaginaries of the reformist state and elites absorbing a new urban lifestyle and a consumer culture in animated places. The creative destruction of the landscape coexists with the socialization of social groups, functionally impacting the subjectivities and consumption practices of gentrifiers.

City branding is an urban developmental strategy often adopted by Chinese local governments. It steers urban change by mobilizing cultural power. Branding as a concept derives from business studies; therefore, the city-branding literature responds to the ways urban management advances the business rationale. This conception implies that cities are increasingly considered commodities that can be customized, promoted and sold [129]. A city brand bridges the otherwise discrete cultural symbols within a locality to create a collective identity of a place or city [35, 162]. This collective identity is then helpful in delineating a unique image of a city, distinguishing the city from others; exhibiting the city to the external; and ultimately, attracting visitors, industries and investment to fuel urban development. Kavaratzis and Ashworth [64] suggested that the process of city branding involves orchestrating activities and narratives to design and create the image of a place or city, which add value to its future. Drawing on the Foucauldian concept of governmentality, Vanolo [129] considered the branding process to be one that creates the subjectification of cities by discursive tactics or the power of knowledge.

Gentrification occurs when the branded identity of the place or city excludes local characters. The danger occurs when the brands of places or cities are profit oriented, aimed at commodifying the places [31]. Additionally, local ruling elites could lead the branding of a declining place to convert it to somewhere safe, clean, and attractive to a new middle class while negating the right of the poor to the city [161]. The specific brand of creative cities that has been favored by many cities has even more directly exposed its target subjects to the creative, talented class and advanced knowledge-based economy [11, 37, 128]. Most commonly, gentrification develops when branding reproduces the symbols and cultural values of other cities, especially those of global or cosmopolitan city status, which causes the standardization of a global city image and the undermining of local identity [35, 68, 102, 129].

Not only the identity of a city or a place, in business studies, city branding is closely tied with the identification of human beings, consumers in particular. City branding takes advantage of the identities of local societies and aims to use them to stand out [8]. Branding in tourist places often highlights the place-based identity of indigenous to include authenticity in the brands. However, some kinds of place-based social identities may be deemed by the brand producers as "negative" and "stigmatic" and therefore something to be abandoned and rebranded [153]. Additionally, branding could shape new identities through guiding, and sometimes normalizing, social behaviors in spaces and defining what is favorable and what is not by discursive tactics [129, 161].

With these connections, Vanolo ([129], p. 102) unravels the political nature of city branding: "the selective storytelling at the basis of branding is connected to the modification of social identities, positionalities and collective images, leading to the production of peculiar forms of inclusion and exclusion." Gentrification studies in China thus need to trace the branding strategies of the state-led market regime in the making of gentrified places and how spatial production and representation attract a group of middle-class consumers and nurture place-based, collective identities.

1.2.2 Spatial Commodification

Departing from the cultural strategy of urban development, much attention has been given to the economic strategy of spatial commodification that propels gentrification. Widespread spatial commodification in Chinese cities can be traced to the land finance strategy of the subnational government. In China, in 1994, nationwide reforms of the tax-sharing system fueled land capitalization and created an emerging phenomenon of land financing in Chinese cities to meet local budgets. The State Council terminated in-kind housing allocations by state-owned enterprises in 1998 and replaced them with workplace subsidies. In this context, urban redevelopment has increasingly departed from self-financing while tending toward the use of social funds and banking loans. Meanwhile, tax reforms have increased the burdens on the local governments responsible for providing urban services. Subnational governments must share half of their revenues with the national government and are responsible for 80% of government expenditures [92]. Local governments have tended to seek land revenues by virtue of the economic responsibilities of local state actors [84, 86, 148]. Regarding the local state in China, Shin Shin (2009) critiqued it for being fully economistic, even though, for example, the government has attempted to increase onsite rehousing for residents who were affected by housing demolition.

Based on this backdrop, an emphasis has been given to the ideological change of the central government in land development and management adjusting to the trajectory of economic transition, and its interactions with multiple actors, such as work units, and national and transnational developers [58, 60, 78, 79, 81–85]. With a city–region horizon, studies have focused on rural–urban land conversion, contextualized by administrative rescaling for rapid urbanization and land-based revenue at the local level. Another stream of the literature investigates the specific practices of institutional arrangements over land-use rights transfer and land appropriation that are often characterized by informal and illegal elements [60, 80, 140, 160].

Unfortunately, few of these studies on institutional changes in land use in a market economy have been contextualized for the case of neighborhood renewal (see [82]). A general institutional arrangement for land reuse at the neighborhood scale follows the procedures of land/property expropriation and compensation, land consolidation and reserve, land auction, planning adjustment and land redevelopment. However, rarely studied are how the sophisticated and conflicting interests in localities have customized these procedures to create specific spatial and social outcomes. This process is typically manifested in taxation reductions and exemptions for land use as ways of encouraging continuing land investment [52]. Less studied is the adjustment of zoning for brownfield redevelopment and the subtle change in housing loans together with policies for housing resettlements. The processes of neighborhood renewal thus provide a microlens through which we can discern the association between the institutional rearrangements of land development, new landscape formation and its social consequences. Such an analysis can illustrate the specific pattern of secondary circuit capital accumulation through investment in the built environment, which has been postulated by Lees et al. [71] as a basic driving force of global contemporary gentrification. As claimed by Ma [89], "since 1990, Chin's urban landscape formation has been deeply affected by how urban land-use rights are transferred and marketed to developers" (p. 1559).

1.2.3 Consumer-Class Formation

While state-led city branding catalyzes the formation of new consumer cultures and collective identities, spatial commodification stimulates new consumption behaviors. Housing is the most important spatial product of property-led, newly built gentrification in China, particularly during the 2000s. Rex and Moore [106] produced a seminal work on housing classes in cities. They argued that the housing market situation, mainly housing tenure access and location, serves for households as a means of articulating political actions in cities. Rex and Moore deemed the main manifestation of class struggles in the city as social units arising around "the control of domestic property" rather than "the use of means of industrial production" (p. 273). The housing class argument has attracted a variety of critiques (see a recent review by [1]. These have denied the housing market position as empirically appropriate for defining social status. However, housing consumption social cleavages, or "consumption sectors" (p. 417) as defined by Dunleavy in 1979, should not be subordinated to or considered to express labor market–induced class divisions. Rather, they should be viewed as an alternative form of social division [107].

The attention paid to housing consumption has directed urban sociologists toward new approaches to the sociology of consumption. Chiefly, authors have paid attention to the characteristics and consumption practices of the newly emerging "service class" ([43], p. 162) along with occupational structural change in postindustrial society compared with the traditional middle class (see also [47, 108]). Given the expansion of the service class defined by Goldthorpe, scholars have raised questions about how and to what extent consumption and lifestyle could increasingly be vehicles leading to class identity (re)formation and middle-class fragmentation (e.g., [13, 75, 108, 135, 145, 163]). The thought trend on middle-class fragmentation and consumption has typically highlighted consumption-side explanations for gentrification.

Unlike the postindustrial context, in China, increasing consumption as a valuable perspective for a class analysis is contingent on housing systemic reform and consumption promotion. Reforms over housing provisions and finance systems during the 1990s paved the way for the new rich to engage in private housing consumption in post-socialist China. Housing consumption facilitates middle-class formation, even though it remains too early to claim an emergent "collective nature of identity" other than property-based collective interests ([125], p. 950, see also [151],

p. 8). Zhang [151] explored the rationales behind an "emerging, fragmented and precarious" middle-class formation in the reform era (p. 6). First, Chinese socialist society lacks a clear-cut middle-class division of lifestyle and identity, upon which one can trace the fragmentation and reformation of classes in the wake of market reform. The contemporary class society in China thus features a co-occurrence of distinction creation and circulation along with the convergence and divergence of social grouping. Thus, occupation may not be an accurate indicator of an individual's wealth because the transitional economy has generated various sources, formally and informally, of affluence. The middle-class cohort in China has been characterized by not only occupational confusion but also, more saliently, the uneven distribution of educational attainment and cultural character (or symbolic capital). The lack of symbolic capital, together with the instability of sources of economic capital and the immature social security system, have all threatened the security of the middle class. As a result, consumption becomes a significant way for individuals to accumulate symbolic capital and declare their social status (see also [28]). Conspicuous consumption practices entrenched in these uniquely social characteristics in the post-socialist period, compared with the latent means of production, have become empirically necessary and viable for a class analysis in China.

In this context, some scholars have underlined two results of private housing consumption on middle-class formation: private housing consumption mobilizes both the cultural (re)production of middle-class distinctions and the spatialization of class distinctions and privileges. Zhou and Chen [157], for instance, argued that, in combination with the collapse of socialist *danwei* institutions, private housing (and transportation) "not only represents consumer goods with which they can build their self-identity and win social recognition but also practice fields for molding new notions of consumption" (p. 94). Employing a top-down perspective, Anagnost [2] deemed the state promotion of consumption practices a type of "national project on a cultural form" (p. 497). The project conforms to the national government's "social engineering" ([117], p. 495, quoted by Anagnost, p. 498) that aims to "expand the middle class by inciting aspiring individuals to adhere to new social norms of middle-class identity often defined around consumer practices" (p. 498).[1] From the consumer perspective, Pow [104] explained that the spatial landscape of gated communities generates a privileged imagination upon which proprietors can proclaim the cultural distinctions of middle-class membership and consolidate the territorial exclusion of cultural otherness. Gated communities become decisive, practical fields forming new patterns of collective interests, mobilizations and their consequential collective conflicts [125, 151].

Therefore, state-led gentrification in China is not merely an outcome of structural social change but also a potential cause. That is, gentrification may not only be

---

[1] This social engineering is conceptualized in China as constructing *xiaokang shehui. Xiaokang shehui* means a "relatively well off" society ([2], p. 502). The term is drawn from the description of an ideal society in the Confucian *Book of Rites* (*Liji*), was first referenced by Deng Xiaoping in 1978 and then recovered by Jiang Zheming authority. Source: Li, S. E. (2003). On Building a Well-off Society in an All Round Way (*Lun quanmian jianshe xiaokang shehui*). *Seek Truth From Facts* (*Shishi qiushi*), 1, 13–16.

induced through the fragmentation of the middle class and the retreat of the working class but also promote class-making through the robust processes of homeownership promotion and place-remaking. According to Lefebvre [73], space is socially produced and carries on the ideologies, values, and meanings of its producers. Space impacts the behaviors and identities of the individuals in it as well as the distribution of power across societies over the right to space. After reconsidering the place and class relationship in gentrification, Butler [15] suggested that *class* studies in gentrification should be broadened beyond the middle and working class struggle over living space to encompass the shaping of class through living-space choices. We are thus encouraged to probe the particularities of the state's urban imaginary and the institutions of spatial commodification in the inner city as well as how state-led spatial practices are then associated with consumer class formation and fragmentation, leading to gentrification. The association between state actors and the new middle class will produce divergent middle-class politics in post-socialist Chinese cities that differ from the liberal gentrifiers found in advanced Western societies.

1.3 Redistribution, Social Governance and Social Re-stratification

Displacement effects are concomitant with gentrification. In China, such effects result from a sophisticated process of state-led spatial resources redistribution and social governance. It impacts the housing provision for low-income residents and the formation of new social inequalities. The discussion integrates to a broad research agenda of social re-stratification under the institutional change in a market economy.

1.3.1 Rehousing Low-Income Residents

Scholars have thoroughly discussed the displacement of the working class (e.g., [1, 4, 5, 17, 136, 143, 144]). They have also intensively criticized studies that tend to neutralize the effects of gentrification. Such studies argue that urban regeneration creates better urban environments and local services not only for the middle class but also for the working class and increases local private and tax revenues that can be used for broad-scale urban services [18, 38, 39]. This position has attracted considerable skepticism, for example, from Newman and Wyly [96]. Other scholars have revisited the concept and methodology of displacement and recovered hidden inequalities in policy-driven gentrification [26, 27, 70, 110]. Since approximately 2000, following the resurgence of policy-driven gentrification in some European and North American cities, scholars interested in class conflicts and the working class have increasingly returned to this focal point to defend the conceptual significance of gentrification. This change has occurred in part because policy discourses such as the urban renaissance, the alleviation of social exclusion, and social mixing have tended to soften public opinion and destabilize to a certain extent the politically critical position held in earlier notions of gentrification [70, 118].

In China, the process of residential relocation in state-facilitated urban redevelopment is commonly known as *chaiqian* (拆迁), which combines the two terms demolition and removal. The process is integral to and regulated by national and municipal governments' policies of land or property expropriation, which involves the agenda of the determination of the subject of expropriation, the object of expropriation and compensation, the compensation methods, and the formulation of resettlement schemes. Essentially, the process creates an opportunity for the housing provision system for low-income residents in the city to be systematically reformed away from the socialist and rural systems and toward a coherent urban housing system.

Because in the old neighborhoods the composition of property rights is complex, the policy-making related to residential relocation has been path-dependent on the current land and housing tenure situations. Shin et al. [115] argued that property rights and tenure relations are key factors complicating gentrification in some East Asian countries. As a result, groupings of current residents may be confronted with different situations and outcomes during and after the process of urban redevelopment and residential relocation (see [82, 113, 114]). For example, residents who are subject to the socialist welfare housing system may have different experiences from those who are already housed by the market system. Those who are residing legally in the city also encounter different housing arrangements from those who are residing illegally in the city (see [113, 114]).

The intricacy of residential relocation thus requires a fine-grained analysis of the different experiences of lower-income residents. Moreover, the existence of "gains" and "losses" among affected residents does not necessarily eliminate the unequal nature of gentrification. Instead, we need to question what kinds of new mechanisms of inequalities are created by gentrification and who suffers the most in this transition.

1.3.2   Social Governance

Decision-making over housing provision for low-income residents affected by gentrification has been greatly impacted by intense state–society interactions. A majority of the existing literature has granted state power predominance in defusing social disagreements in urban redevelopment and gentrification [52, 55, 142, 146]. The system of public ownership of land, sustained since the socialist era, is fundamental to state control over society in the urban redevelopment process. The public ownership of land results in ambiguous private property rights in China's transitional economy, which provides leeway for the Chinese state to arbitrate between the market and society [57, 83]. In policy-making around housing expropriation and relocation, the municipal government has endeavored to define "who gains what rather than who owns what" ([82], p. 884). Through the definition of property rights, for example, the state has tightly managed illegal construction and has retained accountability for the different forms of the rights to live in the city and who possesses them [142]. By introducing the concept of gentrification to describe China's urban change, Shin ([113], p. 484) argued that in an economy without a completely transformed land and housing market, the "dispossession of their [the original residents'] right to properties" by local states acts as a precursor to subsequent investment and gentrification. These studies therefore follow a traditional thesis regarding the totalitarianism of the reformist state in terms of its interactions with lower-income groups [111, 112, 146]. Moreover, the lower-income groups involved in gentrification are all too often treated as one coherent group without internal divisions.

As disagreements arise in the literature, another body of work notes the varied degrees of collaboration in some projects between the market regime and a rising society [24, 56, 152, 158] as well as a few cases of bottom-up regeneration [130]. This change has been witnessed more explicitly in coastal areas, wherein a nascent civil society might be developing, and in the redevelopment projects of urbanized villages, wherein farmers have a strong sense of property rights over collective land. However, rather than collaboration based on equal relationships among stakeholders, government actors often continue to play the dominant role in leading partnership and project operations [56, 158]. Additionally, scholars have emphasized the adaptive side of government policies, such as the increased compensation standards for housing losses and the concessions made to residents who had illegally built housing [61].

In the field of political science, many studies have been concerned with the relationship between the state and society in China after the reform. From the perspectives of social, economic and political life, the discussion extends far from the simplicity of authoritative politics. First, the school of economic elitism based on the strong society theory deems that the capitalists in the generation after the reform will have more decision-making power, and government decision-making will be more subordinate to their interests [42, 77, 88, 122, 138, 139, 147]. These interest groups can even form some embryonic forms of civil society to balance political forces [12, 126, 127]. Second, corporatism, from the perspective of a strong state, implies that the government tends to cooperate with social organizations to resolve the contradictions between communities and improve social control [25, 29, 30, 103]. Tomba [124] explores daily governance in neighborhoods and reveals that the government has constructed an interest structure that intertwines the interests of the state and the community. As a result, daily practices in neighborhoods have internalized the rationality of the regime among residents. Finally, from the perspective of ideological management, Pang [100] considered how the government could maintain the ideological unity of society after reform in even more flexible and decentralized ways.

Among the debates on dualism, some scholars have put forward the concept of mutual empowerment in the state–society relationship [36, 65, 93, 116]. These scholars have refuted state–society theory based on a zero-sum relationship of rights and have argued that the rationality of the state is right based on the effective link between political and social forces. This effective link enables political and social forces to achieve a win–win situation and reach common goals, which finally consolidates the legitimacy of the state. Thus, political forces need social support to achieve collective goals, and society gains its own interests in the process of supporting political forces. This connection leads to what scholars have described as a state-in-society rather than a state-vs.-society relationship. Moreover, as underlined later by Peter Evans, mutually empowering relations are not limited to cooperation between the state and vested interests but include various communities and even vulnerable groups. However, the above views consider only the cooperative relationship between political and social forces, ignoring the suppression and control of political forces on social groups and the internal contradictions of societies.

The implementation of state-led gentrification is a process concerning the conflicts not only between government and communities but also those within communities, relating to both state–society ties and the social control of the state. Understanding the relationship between the government and low-income communities in the process of state-led gentrification is foundational to understanding the process and results of displacement in China.

1.3.3 Social Re-stratification of Low-Income Residents in Post-socialist China

Sociologists in China have raised a debate on Chinese characteristics of institutional change and on how these forces shape forms of social re-stratification in Chinese society. While one view stresses the importance of "incremental change" ([98], p. 108) during institutional transition, the other focuses on "gradual and partial" reform and "evolutionary changes in institutions and policies" ([131], p. 695). These gradual and partial aspects are considered to manifest uniquely in the Chinese case than other transitional economies. While the first emphasizes the cumulative power of non-state actors from the bottom, the latter focuses on the persistence of state actor power.

Nee ([94], with Cao, 1999, with Cao, 2005; with Opper, 2012), a representative sociologist of "market transition theory" (1989, p. 666) focuses on the cumulative market effects of institutional transition. Market transition proponents stress the decisive effects of institutional changes on property rights and opportunities of capital accumulation. They argue that these institutional changes are capable of bringing a "self-reinforcing institutional change" in entrepreneurial capitalism from below ([95], p. 5, 12). These proponents oppose perspectives that unfairly favor monopoly political actors who have arduously defended public ownership, as they argue that such perspectives are ineffective in explaining why and how private markets are growing in China.

Another body of work argues that the persistence of administrative power determines income levels and living quality, although it is accepted that advantages of occupations with market connections are emerging (e.g., [10, 101, 132, 156]). The majority of these studies focus on urban society and the work-units in particular. These authors criticize market transition theory by stating that the proponents of market transition theory assume a convergent outcome of social power dynamics that is ultimately conditioned by market mechanisms, although they do recognize that there are varied forms of path-dependent institutional change among different transitional economies. In China, the "political market" where economic activities operate upon formal and informal ties between state bureaucracy and economic actors, functions saliently insofar as it is susceptible to complete and linear liberal market transition ([101], p. 1043).

So far, a number of works have respectively explored the characteristics of China's land, housing and *hukou* institutional reforms and their influence on social power (e.g., [57, 60, 79, 83, 160]) on land; [87, 134, 159], on housing and *hukou* institutions; [22, 23], on *hukou* and housing demolition and resettlement institutions). However, few of these studies have been contextualized for the case of neighborhood renewal. Processes in gentrification and residential relocation serve as a micro-lens for scrutinizing the transitional and reclassifying processes. They raise questions for the social change of not only the higher-income groups, but, perhaps more intriguingly, the lower-income groups in a transitional economy. The book thus aims to illustrate the power and resource dynamics among difference groups of affected residents in gentrification in detail.

1.4 The Case Study City

Chengdu is selected as a main case study in this thesis. At the meantime, the study approaches the commonalities and differences between Chengdu and other large Chinese cities with prudence, by analysing the socioeconomic and sociocultural backgrounds, urban developmental strategies and gentrification policies comparatively among these cities. Chengdu is a central city of western China and the capital city of Sichuan Province. Since 2000 Chengdu has been at the forefront of the national strategy of "developing China's west"([149], p. 20), which has aimed to transfer the development focus from the eastern and coastal areas to the inland and western areas. Recently, benefiting from the One Belt and Road Initiative promoted by President Jinping Xi, Chengdu's economy has grown sufficiently to make it one of the fastest-developing inland cities in China. During the last two decades, the leadership of Chengdu has advocated various strategies for urban redevelopment, with great ambitions to merge into the global economy. In the meantime, the city is taking advantage of its livable environments, comfortable lifestyle and youthful aesthetics, which has sustained an animated consumer market. In 2008, Chengdu Shanghai, Beijing and Chongqing were selected by global capitalists as among the new worldwide centers of commerce [109]. The state's mobilization of developmental strategies concentrating on the built environment and delightful living conditions appealing to high-income consumers allow this case study of Chengdu to make new claims to gentrification knowledge for the new emerging metropolises in the global economy.

The administrative territory of Chengdu municipality includes traditional urban districts and newly established districts, prefectural-level cities and counties

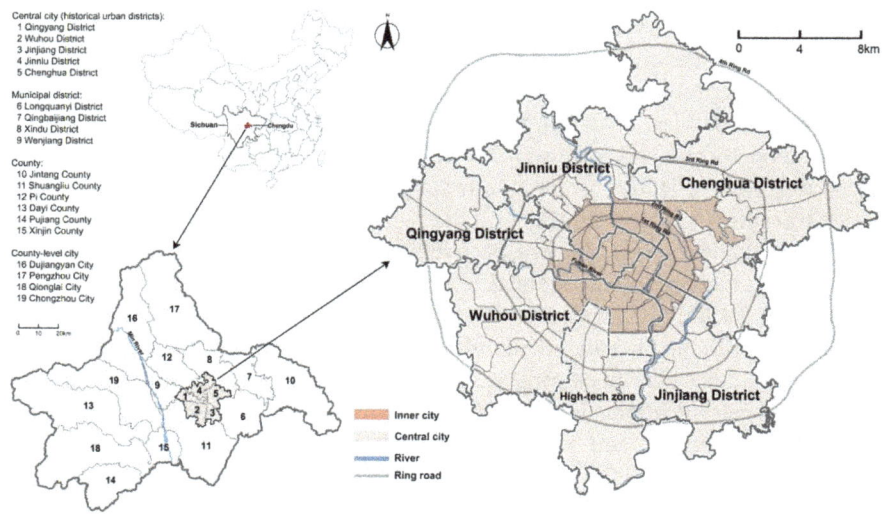

Fig. 1.1 Administrative division of Chengdu municipality and the study area

(Fig. 1.1). The research scope of this study is the central city. It is the most intensively urbanized area in the Chengdu city region and comprises five urban districts
(i.e., Jinjiang, Qingyang, Jinniu, Wuhou and Chenghua) established before 1990. The
gentrification study area is the inner city. Chengdu city is historically a monocentric
city, and the downtown is located close to the geometric center of the metropolitan
region. The inner city of Chengdu is commonly recognized as consisting of the area
within the First Ring Road; essentially, this area was the site of the original city of
Chengdu before the establishment of the People's Republic of China. However, the
modern city of Chengdu has expanded from the original downtown to, currently, the
Fourth Ring Road. This study thus defines the maximum extent of the inner city as the
area within the Second Ring Road (Fig. 1.1). According to the 2010 Chengdu census
and the *2011 Chengdu Statistical Yearbook*, the administrative area of Chengdu was
12,121 km^2 with a resident population of 14 million in 2010, while the five main
urban districts covered 465 km^2 of land and contained 5.03 million inhabitants, and
the inner city was 60 km^2, with a resident population of 1.97 million.

1.5 Methodology

This study took advantage of mixed methods—including statistical and spatial analyses, an institutional analysis, and ethnographic studies—to investigate gentrification
from a structural perspective, the historical perspective, and as a grounded process
within the neighborhood. The triangulation of the three methods and data collection
shape a portrayal of gentrification in Chengdu City from 2000 to 2010.

First, statistical and spatial analyses were used to identify the societal transition related to gentrification and the structural tendencies caused by gentrification at the city level. Specifically, descriptive statistics were used to broadly trace the socioeconomic restructuring and consumer revolution in Chengdu, which generated the broad contexts upon which gentrification occurred. The corresponding social changes in Beijing, Shanghai and Guangzhou were analyzed and compared with Chengdu. Then, the spatial mapping of population change in the census tract showed the geography of gentrification in the inner city. The statistical analysis established correlations between gentrification and the other independent variables.

The above analysis draws on two city-level databases: the three population censuses of Chengdu, Beijing, Shanghai, and Guangzhou in 1990, 2000 and 2010 and the *City Statistical Yearbooks* from 1991 to 2011. The Chinese population census is generated for four administrative divisions that are ranked in descending order: provinces, municipalities, urban districts and urban subdistricts. The last two divisions do not contain rural areas and are often densely inhabited. This study draws on data mainly at the subdistrict level, which is commonly known in China as the street level (*jiedao, 街道*). The geographical area of a subdistrict is larger than that of a neighborhood but smaller than that of an urban district. For the 2010 data, the National Bureau of Statistics keeps the census at the first three levels open source, but the district government, which is below the municipal government, manages the compilation and release of the census at the subdistrict level. I thus collected the data at the subdistrict level by permission of the district governments in the city during the fieldwork.

Second, the study uses an institutional analysis to trace the trajectory of spatial change across waves of state-facilitated urban redevelopment programs. The data for the institutional analysis were drawn from the textual production of the cityscape by the media, policy documents, archives, and official interviews and ethnographic studies on the decision-making of local actors. This study also includes an analysis of urban renewal policies and media reports for Shanghai, Beijing, Guangzhou and other central cities. This part of the analysis helps to elaborate the explanatory power of the Chengdu case study.

Finally, ethnographic studies generate information on the experiences, circumstances and subjective identities of gentrifiers and current residents affected by gentrification at the neighborhood level. This work does not limit the ethnographic studies to individual cases, nor does it aim to tell a single story of neighborhood-based gentrification. We recognize that differences can be found among not only regions but also cases in terms of the driving forces, policy-making and physical modes of gentrification; current residents can experience gentrification and displacement in varied ways. Thus, the study first selected several exemplary case studies (i.e., the Chengdu Paradise, Jintianfu, Caojia Alley (CJA), Jinniu Wanda (JW) and Wide and Narrow Alley (WNA) cases); next, it conducted extensive fieldwork in Chengdu.

1.5.1   Fieldwork in Chengdu

The fieldwork was carried out in three stages. The first round was from December 2, 2014, to April 1, 2015. My primary purpose was to concentrate on the two important cases of CJA and JW redevelopment. Based on a thorough understanding of the operation of the two projects, I could choose the additional field sites to include in the investigation. To investigate the two cases, I interviewed organizational members and relocated residents and collected policy and spatial-planning documents. Additionally, two displacee neighborhoods were identified, where I interviewed migrant tenants who were forced out by the CJA redevelopment. In the latter half of the fieldwork, I started a preliminary investigation of the WNA project, wherein I interviewed relocated residents and retail gentrifiers.

The second round of fieldwork ran from November 2, 2015, to January 31, 2016. The investigation at this stage went beyond the above two cases. Specifically, the WNA project was added as a crucial case because it supplemented the important findings on property and cultural activism that are stressed in this study. Interviews were also conducted in three new neighborhoods where a gentrification process was fully represented and in two new offsite resettlement communities. The identification of residential gentrifiers' neighborhoods was assisted by the spatial mapping of the geography of gentrification in this thesis, which pinpoints the subdistricts that have been most gentrified. In addition, field observations were conducted with residents in an extended geography of redeveloped neighborhoods and resettlement communities in the city to strengthen the findings from the case studies.

The third round of the field work was conducted in the second half of 2019 and the first half of 2020. The first purpose of the last round of the fieldwork was to complement the interviews with the gentrifiers. We selected as a case study a newly built neighborhood (Jintianfu) located in the most gentrified area in Chengdu during the late 2000s. Another 8 gentrifiers were added to our informants. Second, we conducted a questionnaire survey about residential satisfaction and the neighborhood relations of relocated socialist workers to continually trace their changing residential status.

The selection of the field sites in this work was based on three rationales. First, all of the redevelopment projects must be located in the central city (not including resettlement communities located in peripheral areas). Second, these projects together represented different modes of inner-city redevelopment and different methods of compensation to the original residents to allow a relatively complete picture of urban processes and social effects. Third, given that gentrification results in the spatial dispersion of the original residents, cases were selected to provide full coverage of the engaged social groups. Finally, the redevelopment projects under investigation included those involving mainly public/collective properties built by governments and work units and those with a majority prerevolutionary historic buildings.

Table 1.1 lists the basic information about the three key redevelopment projects traced by this study. The first two cases, the *Caojia* Alley (CJA) and *Jinniu Wanda* (JW) projects, concern residential redevelopment. The CJA redevelopment project is

Table 1.1 The profiles of three redevelopment projects in Chengdu

Project name	Completion year	Area (ha)	Redevelopment type	Households receiving compensation			
				Total	Off-site relocation	On-site relocation	Cash
Caojia Alley	2016	13.2	Residential redevelopment of *danwei*/public housing	3756	600	2400	700
Jinniu Wanda	2012	13.6	Residential redevelopment of *danwei*/public housing	3162	162	2100	900
Wide and Narrow Alleys	2008	6.66	Commercial redevelopment of historic buildings	891	891	–	–

Note The total households eligible for claiming compensation include not only owner occupants and public tenants but also absentee owners. They do not contain private tenants who were rejected for compensation

located on the northeast side of the commercial and business centre of Chengdu. Up to four thousand households (14,000 individuals) previously lived in a 13.2 ha area. Except for the rural–urban migrants, the residents are former and current employees of a state-owned construction group called the Huaxi Group and their families. Within the CJA neighborhood, there were 2654 public housing and 880 subsidised units owned by the Huaxi Group and the Jinniu District Housing Department, while another 128 units were commercial apartments.

The CJA redevelopment project is one of the pilot projects of the North Chengdu Redevelopment Programme established by Chengdu municipal government in 2012. The redevelopment project was begun in 2012 and finished at the end of 2016. The government and a financing company led the project as the primary actors in property expropriation and land consolidation; later a private corporation named Evergrande won in the land auction of CJA area and was in charge of real estate development. Impacted by the on-site resettlement policy, among the 3756 households in the old CJA neighborhood who are eligible for compensation, approximately 2400 opted for the state's offer of on-site relocation, while about 600 households were relocated off-site in three communities, which are outside the Third Ring Road of the city. More than 700 households opted for monetary compensation (O2). Aside from the original site of the old neighborhood, three off-site resettlement communities and one displacee area for the original CJA residents were identified in the map. Note that households eligible for claiming compensation include not only owner occupants and public tenants but also absentee owners. This means the 3756 households include residents who were actually living outside the redeveloped neighborhoods when the project was established. And the precise number of private tenants who were rejected for compensation was not recorded.

The second area, Jinniu Wanda (JW) is situated along the north side of the First Ring Road of Chengdu. The 13.6 ha area once accommodated 3162 households. Similar to the CJA neighborhood, the old JW neighborhood also consisted of work unit (*danwei*) compounds, but the *danwei* compounds were almost all small-scale, affiliated with public sectors employers instead of large state-owned enterprises. Also, owing to its proximity to wholesale clothing and houseware markets, the proportion of migrant workers among original residents in the JW neighborhood was larger than the CJA neighborhood. The JW project was from the very beginning endorsed by the government but operated by the Dalian Wanda group, which is commonly deemed to be China's largest private property company. The project started in 2009 and completed in 2012. In this project, 2262 households opted for in-kind compensation, among which 2100 households chose on-site resettlement, whereas only 900 households have opted for monetary compensation or off-site relocation (Jinniu Yearbook, 2012). After redevelopment, the place is separated into three sections. One section is reused for resettling existing residents while the other two are developed respectively to residential and commercial properties for new consumers and investors. Compared with the latter two sections, the resettlement neighborhood is much more densely constructed, containing eight high-rise buildings (more than 2500 apartments) on a 2.2 ha land parcel. The plot ratio reaches to 7.77.

The Wide and Narrow Alleys (WNA) was previously a *hutong* (alley, or narrow street) neighborhood adjacent to the west side of the CBD of Chengdu. The 6.6 ha area contained three *hutong*s and courtyard dwellings that were constructed in late imperial China and the pre-revolutionary period. Initially, the WNA area was constructed as a station of Manchu military power in Chengdu in the Qin Dynasty. The area was managed by the military power beyond the jurisdiction of Chengdu administration. Thus, the WNA is commonly known as *Shaocheng* (small city), which means a small city nested independently within the *Dacheng* (big city) of Chengdu. Before redevelopment, the place was inhabited by local residents, who were in possession of the buildings in the pre-revolutionary days, and public tenants, who moved in after the buildings were confiscated by the government and shifted to public housing. From 2003 to 2008, the WNA area experienced commercial redevelopment. Most of the traditional buildings were demolished and reconstructed using traditional construction techniques. The area is now a renowned tourist place in Chengdu. The project caused wholesale off-site relocation of local inhabitants. All of the 891 indigenous residents were relocated off-site during the process of commercialisation.

Three major groups participated in the study: organizational members (6 city officials, 2 *danwei* managers and 1 manager of the state-owned development company in the CJA project, and 6 urban planners and architects who managed the operation of the JW and CJA projects), relocated residents who had originally lived in the old neighborhoods and have either been relocated into new communities or are still waiting for relocation (20 offsite, 20 onsite, 10 activists, and 20 displacees), and gentrifiers (20 residential gentrifiers from four neighborhoods in the inner city). The activists were residents who intended to remain and who had intensely resisted the housing demolition and removal. In addition to the interviews, two focus groups were organized with residents and experts. In the subsequent text, each interviewee is identified by a classification code—O for the organizational members, R for the relocated residents, and G for the gentrifiers—followed by a chronological interview number.

1.6 Overview

This book provides a thematic analysis of the stages of gentrification in the case study city, emphasizing the interactional influences among the actors. Part One generally sketches the contexts, manifestations and relevance of gentrification in a Chinese city. It starts with the context for socioeconomic restructuring and sociocultural change in post-reform China (Chap. 1). Then, Chap. 2 maps the geography of the socio-spatial upgrade in the inner city of Chengdu in the 2000s to confirm the extent and spatial manifestation of gentrification. It then correlates a social upgrading index with a series of independent variables, aiming to establish a dependent relationship between the gentrification process and broader societal and spatial transformation in the post-socialist era.

Departing from structural tendencies, Part Two examines the process of inner-city upgrading, considering the agency of both politico-economic actors and middle-class consumers in the city. The role of the state is emphasized, particularly state–capital–society relations, thereby linking state intervention with societal change. Taking a historical perspective and using an institutional analysis, Chap. 3 examines two aspects of the urban redevelopment practices conducted by politico-economic actors: cultural place-making and spatial commodification. These two practices direct the study toward the implications for social dynamics—the construction of new consumer cultures and the elevation of spatial consumption—that underlie the genesis of gentrification in China. Chapter 4 then analyzes the cultural and identity changes of middle-class gentrifiers and explains the causes and patterns of the gentrifying process by combining both production and consumption forces.

Part Three turns to the experience of indigenous low-income residents in gentri-fication. Chapter 5 introduces the changing housing and compensation policies related to the property acquisition of local residents affected by urban redevelopment projects. Then, through ethnographic studies, the next three chapters investigate the circumstances, attitudes and outcomes of three groups of indigenous residents during the process of residential relocation induced by inner-city redevelopment, highlighting the diverse state–society interactions in gentrification. The state–society interactions of cooperation, confrontation and submission are explored to raise a broader discussion about social outcomes along with gentrification and relocation.

The concluding chapter brings together key findings of this study. The book generalizes the three aspects of institutional changes in the cultural, economic and social spheres that have thus far directed the operation of gentrification in Chengdu: the creative destruction of consumption spaces, the spatial production of excess, and the unequal redistribution of spatial resources to low-income residents. These three aspects of institutional changes characterize gentrification in Chengdu as a transformative force of economic and social development. It is led by the state and capitalists and endorsed by middle-class consumers. Four regional characteristics of gentrification in Chengdu are revealed. The study suggests that flexible, pragmatic and diverse state–society relations explain the formation of these regional peculiarities. Finally, the spatiotemporality of argumentations in this study is discussed by comparing the Chengdu case study with situations in other Chinese metropolis and with the new tendencies of urban renewal they have shown.

References

1. Allen C (2008) Gentrification "research" and the academic nobility: a different class? Int J Urban Reg Res 32(1):180–185
2. Anagnost A (2008) From "class" to "social strata": grasping the social totality in reform era China. Third World Q 29(3):497–519
3. Appadurai A (1990) Disjuncture and difference in the global cultural economy. Publ Cult 2(2):1–24
4. Atkinson R (2000) The hidden costs of gentrification: displacement in central London. J Housing Built Environ 15(4):307–326
5. Atkinson R (2002) Does gentrification help or harm urban neighborhoods? Centre for Neighborhood Research, Glasgow
6. Atkinson R, Bridge G (2005) Introduction. In: Atkinson R, Bridge G (eds) Gentrification in a global context: the new urban colonialism. Routledge, London, pp 1–17
7. Badyina A, Golubchikov O (2005) Gentrification in central Moscow: a market process or a deliberate policy? Money, power and people in housing regeneration in Ostozhenka. Geografiska Annaler B 87(2):113–129
8. Bailey C, Miles S, Stark P (2004) Culture-led urban regeneration and the revitalisation of identities in Newcastle, Gateshead and the North East of England. Int J Cult Policy 10(1):47–65
9. Betancur JJ (2014) Gentrification in Latin America: overview and critical analysis. Hindawi Publishing Corporation Urban Studies Research. Retrieved 23 Nov 2015, from: https://doi.org/10.1155/2014/986961
10. Bian YJ, Logan JR (1996) Market transition and the persistence of power: the changing stratification system in urban China. Am Sociol Rev 61(5):739–758
11. Bianchini F (1993) Culture, conflict and cities: issues and prospects for the 1990s. Cultural policy and urban regeneration: the West European experience, pp 199–213
12. Brook T, Frolic BM (1997) Civil society in China. ME Sharpe, New York
13. Butler T (1997) Gentrification and the middle classes. Ashgate, Aldershot
14. Butler T (2003) Living in the bubble: gentrification and its' others' in North London. Urban Stud 40(12):2469–2486
15. Butler T (2007) For gentrification? Environ Plan A 39(1):162–181
16. Butler T, Lees L (2006) Super-gentrification in Barnsbury, London: globalisation and gentrifying global elites at the neighborhood level. Trans Inst Br Geogr 31(4):1–21
17. Butler T, Hamnett C, Ramsden M (2013) Gentrification, education and exclusionary displacement in east London. Int J Urban Reg Res 37(2):556–575
18. Byrne JP (2003) Two cheers for gentrification. Harvard Law J 46(3):405–432
19. Cartier C (2009) Production/consumption and the Chinese city/region: cultural political economy and the feminist diamond ring. Urban Geogr 30(4):368–390
20. Cartier C (2013) Building civilised cities. In: Barmé G, Goldkorn J (eds) China story yearbook 2013: civilising China. The Australian National University, Canberra, pp 256–285

21. Cartier C (2016) Governmentality and the urban economy: consumption, excess, and the 'civilised city' in China. In: Bray D, Jeffreys E (eds) New mentalities of government in China. Routledge, New York, pp 56–73
22. Chen YF (2008) The crisis of legitimacy and space of rationality in urban development (Chengshi kaifa de zhengdangxing weiji yu helixing kongjian). Sociol Stud (shehuixue yanjiu) 3:29–55 (in Chinese)
23. Chen YF (2009) City development and housing exclusion: the true meaning of city access system (chengshikaifa yu zhufangpaici: chengshizhunruzhi de biaoxiang yu shizhi). J Ningbo Univ (Liberal Arts Edition) (ningbo daxue xuebao renwen kexue ban) 2(22)::32–39 (in Chinese)
24. Chen YW, Qu L (2020) Emerging participative approaches for urban regeneration in Chinese megacities. J Urban Plan Dev 46(1):04019029
25. Cheng L (ed) (2009) China's changing political landscape: prospects for democracy. Brookings Institution Press
26. Davidson M (2008) Spoiled mixture: where does state-led "positive" gentrification end? Urban Stud 45(12):2385–2406
27. Davidson M (2009) Displacement, space and dwelling: placing gentrification debate. Ethics Place Environ 12(2):219–234
28. Davis D (2005) Urban consumer culture. China Q 183:692–709
29. Dickson B (2003) Red Capitalists in China: the party, private entrepreneurs and prospects for political change. Cambridge University Press, Cambridge
30. Dickson BJ (2008) Wealth into power: the Communist Party's embrace of China's private sector. Cambridge University Press, Cambridge
31. Dinardi C (2017) Cities for sale: contesting city branding and cultural policies in Buenos Aires. Urban Stud 54(1):85–101
32. Doshi S (2015) Rethinking gentrification in India: displacement, dispossession and the spectre of development. In: Lees L, Shin HB, López-Morales E (eds) Global gentrifications: uneven development and displacement. Policy Press, Bristol, UK, pp 101–120
33. Dowall DE (1993) Urban redevelopment in the People's Republic of China. Housing Finance Int 7(3):25–35
34. Elshahed M (2015) The prospects of gentrification in downtown Cairo: artists, private investment and neglectful state. In: Lees L, Shin HB, López-Morales E (eds) Global gentrifications: uneven development and displacement. Policy Press, Bristol, UK, pp 121–142
35. Evans G (2003) Hard-branding the cultural city—from Prado to Prada. Int J Urban Reg Res 27(2):417–440
36. Evans P (1995) Embedded autonomy: states and industrial transformation. Princeton University Press, Princeton
37. Florida R (2002) The rise of the creative class. Basic Books, New York
38. Freeman L (2006) There goes the hood: views of gentrification from the ground up. Temple University Press, Philadelphia
39. Freeman L, Braconi F (2004) Gentrification and displacement: New York City in the 1990s. J Am Plann Assoc 70(1):39–52
40. Ghertner DA (2014) India's urban revolution: geographies of displacement beyond gentrification. Environ Plan A 46:1554–1571
41. Ghertner DA (2015) Why gentrification theory fails in "much of the world." City 19(4):552–563
42. Gold TB (1990) Party-state versus society in China. In: Kallgren JK (ed) Building a nation-state: China at forty. Institute of East Asian Studies, University of California, Berkeley, pp 125–151
43. Goldthorpe JH (1982) On the service class: its formation and future. In: Giddens A, MacKenzie G (eds) Social class and the division of labour. Macmillan, Basingstoke
44. Hackworth J (2007) The neoliberal city: governance, ideology and development in American urbanism. Cornell University Press, Ithaca, New York

45. Hackworth J, Smith N (2001) The changing state of gentrification. Tijdschr Econ Soc Geogr 92(4):464–477
46. Hall PA, Taylor CR (1996) Political science and the three new institutionalisms. Polit Stud 44(5):936–957
47. Hamnett C (1989) Consumption and class in contemporary Britain. In: Hamnet C, McDowell L, Sarre P (eds) Restructuring Britain: the changing social structure. Sage, London
48. Hamnett C (2020) Is Chinese urbanisation unique? Urban Stud 57(3):690–700
49. Harris A (2008) From London to Mumbai and back again: gentrification and public policy in comparative perspective. Urban Stud 45(12):2407–2428
50. Harvey D (1989) From managerialism to entrepreneurialism: the transformation in urban governance in late capitalism. Geografiska Annaler Series B Human Geography 71(1):3–17
51. Harvey D (1990) The condition of postmodernity: an enquiry into the origins of cultural change. Blackwell, Cambridge, MA
52. He S (2007) State-sponsored gentrification under market transition: the case of Shanghai. Urban Aff Rev 43(2):171–198
53. He S (2010) New-build gentrification in central Shanghai: demographic changes and socioeconomic implications. Popul Space Place 16(5):345–361
54. He S, Wu F (2005) Property-led redevelopment in post-reform China: a case study of Xintiandi redevelopment project in Shanghai. J Urban Aff 27(1):1–23
55. He S, Wu F (2009) China's emerging neoliberal urbanism: perspectives from urban redevelopment. Antipode 41(2):282–304
56. Hin LL, Xin L (2011) Redevelopment of urban villages in Shenzhen, China—an analysis of power relations and urban coalitions. Habitat Int 35(3):426–434
57. Ho P (2001) Who owns China's land? Policies, property rights and deliberate institutional ambiguity. China Q 166:394–421
58. Ho SPS, Lin GCS (2003) Emerging land markets in rural and urban China: policies and practices. China Q 175:681–707
59. Hsing Y (2006) Land and territorial politics in urban China. China Q 187:1–18
60. Hsing Y (2010) The great urban transformation: politics of land and property in China. Oxford University Press, New York
61. Hu FZ, Lin GC, Yeh AG, He S, Liu X (2020) Reluctant policy innovation through profit concession and informality tolerance: a strategic relational view of policy entrepreneurship in China's urban redevelopment. Public Adm Dev 40(1):65–75
62. Huang X, Liu Y (2021) Worse life, psychological suffering, but 'better' housing: the post-gentrification experiences of displaced residents from Xuanwumen, Beijing. Popul Space Place e2523
63. Iossifova D (2009) Negotiating livelihoods in a city of difference: narratives of gentrification in Shanghai. Crit Plan 16:98–116
64. Kavaratzis M, Ashworth GJ (2005) City branding: an effective assertion of identity or a transitory marketing trick? Tijdschr Econ Soc Geogr 96(5):506–514
65. Kohli A (2004) State-directed development: political power and industrialization in the global periphery. Cambridge University Press
66. Kovács Z, Wiessner R, Zischner R (2013) Urban renewal in the inner city of Budapest: gentrification from a post-socialist perspective. Urban Stud 50(1):22–38
67. Kyung S, Kim K (2011) State-facilitated gentrification in Seoul, South Korea: for whom, by whom and with what result? Paper presented at the International RC21 conference, Amsterdam
68. Landry C, Bianchini F (1995) The Creative City. Demos, London
69. Lees L (2003) Super-gentrification: the case of Brooklyn Heights New York City. Urban Stud 40(12):2487–2509
70. Lees L (2008) Gentrification and social mixing: towards an inclusive urban renaissance. Urban Stud 45(12):2449–2470
71. Lees L, Shin HB, López-Morales E (2016) Planetary gentrification. Polity Press, Cambridge, Malden, MA
72. Lees L, Slater T, Wyly E (2008) Gentrification. Routledge, New York

73. Lefebvre H (1991) The production of space (Nicholson-Smith D, trans). Basil Blackwell, Oxford
74. Lemanski C (2014) Hybrid gentrification in South Africa: theorising across southern and northern cities. Urban Stud 51(14):2943–2960
75. Ley D (1996) The new middle class and the remaking of the central city. Oxford University Press, Oxford
76. Ley D, Teo SY (2014) Gentrification in Hong Kong? Epistemology vs. ontology. Int J Urban Reg Res 38(4):1286–1303
77. Li C (2010) China's emerging middle class: beyond economic transformation. Brookings Institution Press, Washington, DC
78. Lin GCS (2007) Reproducing spaces of Chinese urbanisation: new city-based and land-centred urban transformation. Urban Stud 44(9):1827–1855
79. Lin GCS (2009) Developing China: land, politics, and social conditions. Routledge, London
80. Lin GCS (2010) Understanding land development problems in globalising China. Eurasian Geogr Econ 51(1):80–103
81. Lin GCS (2014) China's landed urbanisation: neoliberalising politics, land commodification, and municipal finance in the growth of metropolises. Environ Plan A 46:1814–1835
82. Lin GCS (2015) The redevelopment of China's construction land: practising land property rights in cities through renewals. China Q 224:865–887
83. Lin GCS, Ho SPS (2005) The state, land system, and land development processes in contemporary China. Ann Assoc Am Geogr 95(2):411–436
84. Lin GCS, Yi F (2011) Urbanisation of capital or capitalisation on urban land? Land development and local public finance in urbanising China. Urban Geogr 32:50–79
85. Lin GCS, Zhang AY (2015) Emerging spaces of neoliberal urbanism in China: land commodification, municipal finance and local economic growth in prefecture-level cities. Urban Stud 52(15):2774–2798
86. Liu T, Lin GCS (2014) New geography of land commodification in Chinese cities: uneven landscape of urban land development under market reforms and globalisation. Appl Geogr 51:118–130
87. Logan JR, Bian YJ, Bian FQ (1999) Housing inequality in urban China in the 1990s. Int J Urban Reg Res 23(1):7–25
88. Lu XY (2002) Zhongguo Shehui Jieceng Baogao (Social stratification in contemporary China). Social Science Academic Press, Beijing
89. Ma LJC (2002) Urban transformation in China, 1949–2000: a review and research agenda. Environ Plan A 34:1545–1569
90. Maloutas T (2007) Segregation, social polarisation and immigration in Athens during the 1990s: theoretical expectations and contextual difference. Int J Urban Reg Res 31(4):733–758
91. Maloutas T (2014) Social and spatial impact of the crisis in Athens—from clientelist regulation to sovereign debt crisis. Region et Developpement 39:149–166
92. Man JY (2010) Local public finance in China: an overview. In: Man JY, Hong Y-H (eds) China's local public finance in transition. Lincoln Institute of Land Policy, Cambridge, MA, pp 3–20
93. Migdal J (2001) State in society: studying how states and societies transform and constitute one another. Cambridge University Press, Cambridge
94. Nee V (1989) A theory of market transition: from redistribution to markets in state socialism. Am Sociol Rev 54(5):663–681
95. Nee V, Opper S (2012) Capitalism from below: markets and institutional change in China. Harvard University Press, Cambridge, MA
96. Newman K, Wyly EK (2006) The right to stay put: revisited gentrification and resistance to displacement in New York City. Urban Stud 43(1):23–57
97. Ning Y, Chang TC (2021) Production and consumption of gentrification aesthetics in Shanghai"s M50. Trans Inst Br Geogr
98. North D (1991) Institutions. J Econ Perspect 5(1):97–112

99. Ong A (1997) Chinese modernities: narratives of nation and of capitalism. Ungrounded empires: the cultural politics of modern Chinese transnationalism, pp 171–202
100. Pang Q (2018) State-society relations and Confucian revivalism in contemporary China. Springer
101. Parish WL, Michelson E (1996) Politics and markets: dual transformations. Am J Sociol 101(4):1042–1059
102. Peck J (2005) Struggling with the creative class. Int J Urban Reg Res 29(4):740–770
103. Pei MX (2006) China's trapped transition: the limits of developmental autocracy. Harvard University Press, Cambridge, MA
104. Pow CP (2009) Gated communities in China: class, privilege and the moral politics of the good life. Routledge, London
105. Ren J (2015) Gentrification in China? In: Lees L, Shin HB, López-Morales E (eds) Global gentrifications: uneven development and displacement. Policy Press, Bristol, UK, pp 329–348
106. Rex J, Moore R (1967) Race, community and conflict. Oxford University Press, Oxford
107. Saunders P (1984) Beyond housing classes: the sociological significance of private property rights in means of consumption. Int J Urban Reg Res 8(2):202–227
108. Savage M, Barlow J, Dickens P, Fielding T (1992) Property, bureaucracy and culture: middle-class formation in contemporary Britain. Routledge, London
109. Scott AJ (2011) Emerging cities of the third wave. City 15(3–4):289–321
110. Shaw K, Hagemans I (2015) "Gentrification without displacement" and the consequent loss of place: the effects of class transition on low-income residents of secure housing in gentrifying areas. Int J Urban Reg Res 39(2):329–341
111. Shin HB (2009) Life in the shadow of mega-events: Beijing Summer Olympiad and its impact on housing. J Asian Public Policy 2(2):122–141
112. Shin HB (2009) Residential redevelopment and the entrepreneurial Local State: the implications of Beijing's shifting emphasis on urban redevelopment policies. Urban Stud 46(13):2815–2839
113. Shin HB (2016) Economic transition and speculative urbanisation in China: gentrification versus dispossession. Urban Stud 53(3):471–489
114. Shin HB, Li B (2013) Whose games? The costs of being "Olympic citizens" in Beijing. Environ Urban 25(2):559–576
115. Shin HB, Lees L, López-Morales E (2016) Introduction: locating gentrification in the Global East. Urban Stud 53(3):455–470
116. Shue V (1988) The reach of the state: sketches of the Chinese body politics. Stanford University Press, Stanford, CA
117. Sigley G (2006) Chinese governmentalities: government, governance and the socialist market. Economy Economy Soc 35(4):487–508
118. Slater T (2006) The eviction of critical perspectives from gentrification research. Int J Urban Reg Res 30(4):735–757
119. Smith N (2002) New globalism, new urbanism: gentrification as global urban strategy. Antipode 34(3):427–450
120. Smith N, Defilippis J (1999) The reassertion of economics: 1990s gentrification in the Lower East Side. Int J Urban Reg Res 23(4):638–653
121. Steinmo S (2008) What is historical institutionalism? In: Della Porta D, Keating M (eds) Approaches and methodologies in the social sciences. Cambridge University Press, Cambridge, UK, pp 118–138
122. Strand D (1990) Protest in Beijing: civil society and public sphere in China. Problems of communism, vol 39, pp 1–19
123. Sýkora L (2005) Gentrification in post-communist cities. In: Atkinson R, Bridge G (eds) Gentrification in a global context: the new urban colonialism. Psychology Press, New York, pp 99–105
124. Tomba L (2014) The government next door. Cornell University Press
125. Tomba L (2005) Residential space and collective interest formation in Beijing's housing disputes. China Q 184(1):934–951

126. Unger J, Chan A (1995) China, corporatism, and the East Asian model. Aust J Chin Aff 33:29–53
127. Unger J, Chan A (1996) Corporatism in China; a developmental state in an East Asian context. In: McCormick B, Unger J (eds) China after socialism. In the footsteps of Eastern Europe or East Asia? M.E. Sharpe, Armonk, NY, pp 95–129
128. Vanolo A (2008) The image of the creative city: some reflections on urban branding in Turin. Cities 25(6):370–382
129. Vanolo A (2017) City branding: the ghostly politics of representation in globalising cities. Routledge
130. Verdini G (2015) Is the incipient Chinese civil society playing a role in regenerating historic urban areas? Evidence from Nanjing, Suzhou and Shanghai. Habitat Int 50:366–372
131. Walder AG (1995) China's transitional economy: interpreting its significance. China Q 144(Special Issue: China's Transitional Economy):963–979
132. Walder AG (1996) Markets and inequality in transitional economies: toward testable theories. Am J Sociol 101(4):1060–1073
133. Wang J, Lau S (2009) Gentrification and Shanghai's new middle-class: another reflection on the cultural consumption thesis. Cities 26(2):57–66
134. Wang Y, Murie A (2000) Social and spatial implications of housing reform in China. Int J Urban Reg Res 24(2):397–417
135. Warde A (1994) Consumption, identity-formation and uncertainty. Sociology 28(4):877–898
136. Watt P (2008) The only class in town? Gentrification and the middle-class colonisation of the city and the urban imagination. Int J Urban Reg Res 32(1):206–211
137. Weinstein L, Ren X (2009) The changing right to the city: urban renewal and housing rights in globalising Shanghai and Mumbai. City Community 8(4):407–432
138. White G (1993) Riding the tiger: the politics of economic reform in post-Mao China. Stanford University Press, Stanford, CA
139. Whyte MK (1992) Prospects for democratization in China. Problems of communism, pp 58–70
140. Wong KK, Zhao XB (1999) The influence of bureaucratic behaviour on land apportionment in China: The informal process. Environment and Planning C 17:113–126
141. Wu F (2004) Residential relocation under market-oriented redevelopment: The process and outcomes in urban China. Geoforum 35(4):453–470
142. Wu F (2016) State dominance in urban redevelopment: beyond gentrification in urban China. Urban Aff Rev 52(5):631–658
143. Wyly EK, Newman K, Schafran A, Lee E (2010) Displacing New York. Environ Plan A 42:2602–2623
144. Wyly E, Hammel D (2004) Gentrification, segregation, and discrimination in the American urban system. Environ Plan A 36(7):1215–1241
145. Wynne D (1998) Leisure, lifestyle and the new middle class. Routledge, London
146. Yang YR, Chang CH (2007) An urban regeneration regime in china: a case study of urban redevelopment in Shanghai's Taipingqiao area. Urban Stud 44(9):1809–1826
147. Yang MM (1989) Between state and society: the construction of corporateness in a Chinese socialist factory. Aust J Chin Aff 22(July):31–60
148. Yeh AGO, Wu F (1996) The new land development process and urban development in Chinese cities. Int J Urban Reg Res 20(2):330–353
149. Yeung YM (2004) Introduction. In Yeung YM, Shen J (eds) Developing China's west: a critical path to balanced national development. The Chinese University of Hong Kong, Hong Kong
150. Yip NM, Tran HA (2016) Is "gentrification" an analytically useful concept for Vietnam? A case study of Hanoi. Urban Stud 53(3):490–505
151. Zhang L (2010) In search of paradise: middle-class living in a Chinese metropolis. Cornell University Press, New York
152. Zhang L, Chen J, Tochen MR (2016) Shifts in governance modes in urban redevelopment: a case study of Beijing's Jiuxianqiao Area. Cities 53:61–69

153. Zhang Q, Lu L, Huang J, Zhang X (2021) Uneven development and tourism gentrification in the metropolitan fringe: a case study of Wuzhen Xizha in Zhejiang Province, China. Cities 103476
154. Zhang X, Hu J, Skitmore M, Leung BYP (2014) Inner-city urban redevelopment in China metropolises and the emergence of gentrification: case of Yuexiu, Guangzhou. J Urban Plan Dev 140(4):1–8
155. Zhang Y, Fang K (2004) Is history repeating itself?: From urban renewal in the United States to inner city redevelopment in China. J Plan Educ Res 23(3):286–298
156. Zhou M, Logan JR (1996) Market transition and the commodification of housing in urban China. Int J Urban Reg Res 20(3):400–421
157. Zhou XH, Chen Q (2010) Globalisation, social transformation, and the construction of China's middle class. In: Li C (ed) China's emerging middle class. Brookings Institution, Washington, pp 84–103
158. Zhou ZH (2014) Towards collaborative approach? Investigating the regeneration of urban village in Guangzhou, China. Habitat Int 44:297–305
159. Zhu JM (2000) The changing mode of housing provision in transitional China. Urban Aff Rev 35(4):502–519
160. Zhu JM (2004) From land use right to land development right: Institutional change in China's urban development. Urban Studies 41(7):1249–1267
161. Zukin S (1982) Loft living: culture and capital in urban change. The Johns Hopkins University Press, Baltimore
162. Zukin S (1995) The cultures of cities. Blackwell, Oxford and Cambridge, MA
163. Zukin S (1998) Urban lifestyles: diversity and standardisation in spaces of consumption. Urban Stud 35(5–6):825–839

Part I
Foundation and Manifestation

Chapter 2
Social Transformation in Large Chinese Cities

Although various explanations have been proffered, the transformation to a postindustrial society is generally treated by gentrification scholars as an important backdrop to gentrification. In the 1960s and 1970s, following the arrival of urban pioneers, who tended to renovate inner-city landscapes and lifestyles (see [13, 36]), the expansion of high-status service classes—specifically professional, technical and managerial workers—created a pool of potential consumers for inner-city housing and lifestyles. The contexts of postindustrial transformation and middle-class consumer culture demand reexamination in contemporary large cities in China. Two aspects of the background are highlighted: socioeconomic restructuring in the new economy, which is characterized by the primary stage of developing knowledge-based services, and the rise of the new rich and the reintroduction of individual consumption into personal life, which prompts the formation of middle-class cultural distinctions in a post-socialist society.

2.1 Socioeconomic Restructuring in the New Economy

The rise of major cities in China as new destinations of transnational service corporations has been well documented. Edgington and Haga [7] found an obvious expansion in the number of Japan-based service companies in Beijing and Shanghai from 1985 to 1995, although the numbers in Guangzhou and Shenzhen did not increase proportionally. In an attempt to associate capitalist economic development with the urbanization process worldwide, Scott [25] collected materials on emerging third wave cities following the movement of capitalism toward "a global cognitive–cultural economy" (p. 295). These cities have been well regarded by global capitalists for their business potential. The Chinese cities of Shanghai, Beijing, Chengdu and Chongqing (the latter two are inland cities in Western China) are on the list and are marked

---

Parts of the findings from this chapter have been published in Geoforum (2018, 93, pp. 120–132).

© Springer Nature Singapore Pte Ltd. 2022
Q. Yang, *Gentrification in Chinese Cities*, Urban Sustainability,
https://doi.org/10.1007/978-981-19-2286-2_2

as cities in transformation from a previously "marginal" status to among the new global high fliers. Nevertheless, the transformation toward a quaternary economy is incomplete, and these cities remain dominated by tertiary industries.

Based on the population censuses of 1990, 1995, 2000 and 2010, this study traces the labor-restructuring in four typical large cities in China: Beijing, Shanghai, Guangzhou and Chengdu.[1] While Beijing is the capital city of China, Shanghai is commonly recognized as the nation's most advanced city economy. Guangzhou and Chengdu, on the other hand, are representative provincial capitals. For the four cities, datasets are generated on the industrial and occupational populations for the main urban districts.[2] In China, the administrative definition of a metropolitan region includes traditional urban and newly established districts as well as prefectural-level cities and counties. By confining the analysis to the main urban districts, this study emphasizes employment restructuring in intensively urbanized areas within a city–region. Although the definition is not unified, main urban districts generally refer to the commonly recognized traditional districts established at least before the 1990s.

2.1.1 Changing Industrial Structure

Employment categorizations vary with each census. In particular, market reform in China has transformed certain sectors from public to private. For example, real estate management was combined with public and residential services in 1990 but was classified separately in 2000 and 2010. Meanwhile, new sectors such as leasing and business services were established in 2010, and censuses may combine sectors listed separately in prior census data. In this case, the data of the employed population are re-sorted into a unified classification, thereby allowing comparisons between censuses. In particular, the service sector is divided into distributive, personal, producer and social services and public administration. Distributive services include transport, storage, and postal services; information and communication; and whole-sale and retail trade. Personal services combine the industries of accommodation and catering; culture, sports, and entertainment; and residential, repair and other services. Producer services aggregate finance, insurance and real estate (FIRE) with business and professional services. Social and public services contain sectors in education,

---

[1] The census data for the occupational and industrial population in 1990 were based on the total population, whereas the 2000 and 2010 census data were both based on a 10% sampling of the total population. In this study, the 2000 and 2010 data have been multiplied by ten to allow estimation of the true populations in those two years.

[2] The main urban districts of Chengdu include Jinjiang, Qingyang, Jinniu, Wuhou and Chenghua districts. The main urban districts of Shanghai include Huangpu, Luwan, Xuhui, Changning, Jingan, Putuo, Zhabei, Hongkou and Yangpu districts. The main urban districts of Beijing include Dongcheng, Xicheng, Chongwen, Xuanwu, Chaoyang, Fengtai, Shijingshan, Haidian. The main urban districts of Guangzhou include Dongsan, Liwan, Yuexiu, Haizhu, Tianhe, Fangcun.

health care, social insurance and welfare along with public administration and social organizations.[3]

From 1990 to 2010, the four city economies transitioned from manufacturing to service industries (Fig. 2.1). The number of those employed in the manufacturing sectors declined from 54.44% of the employed population in Shanghai in 1990 to 17.40% in 2010. The changes in Guangzhou and Chengdu were smaller, from 35.48% to 21.13% and from 35.35% to 15.40%, respectively. In Beijing, because the manufacturing industry was previously relocated from the central urban area, the proportion of the manufacturing industry dropped to 27.19% in 1995 and continued to decline over the next 15 years to 10.04%. In contrast to the changeover seen in manufacturing, service labor increased in all sectors except for social services in the four cities. By 2010, distributive services captured the largest percentage among the four types of the service sector, amounting to more than one-third of total employment. Although producer services achieved the highest growth rate among all service industries across the twenty years studied, the labor force in these high-skilled industries accounted for less than one-fifth of employment in the four cities in 2010. In Shanghai (16.38%) and Beijing (17.90%), the numbers employed in producer services exceeded those of personal and social services in 2010, while in that same year, producer service employment (11.52%) remained slightly lower than personal service employment (12.97%) in Chengdu. For Guangzhou, the proportions of the three service sectors were approximately equal. Compared with [9, 10] observation of Greater London in 1998, when financial and business service workers accounted for one-third of all employees, their counterparts in the large cities in China were still much fewer in number in 2010. According to [24], the FIRE group represented 29.8% of the total employment in New York City in 1977 and 28% of that in London in 1971 (p. 132). In the four Chinese cities, this group accounted for less than 10% of the employed. Hence, despite showing strong momentum, knowledge-based service industrial jobs remained of modest size in large Chinese cities.

In the five main urban districts of Chengdu, employment almost doubled from 1990 to 2010 (Table 2.1). However, the ratio between the employed population and the population aged 15 and over actually declined, from 71.4% in 1990 to 57.7% in 2000 and to 54.9% in 2010. In 2010, the five traditional urban districts of Chengdu were inhabited by more than 4.5 million people aged 15 years and over, of which the employed population was 2.5 million. The increasing unemployment rate can be explained by referring to the labor market reform during China's transitional period. Based on the labor allocation system designed in the planned economy, local governments attempted to guarantee full employment through the straightforward assignment of students to state sectors after graduation. The labor system–generated redundant labor and low labor mobility ultimately reduced firm productivity [2]. In accordance with enterprise reform, the labor market reform enabled managers to gain

---

[3] The four categories are based on the classification of the Organization of Economic Cooperation and Development; the minor categories within the four sections are based on the International Standard of Industrial Classification.

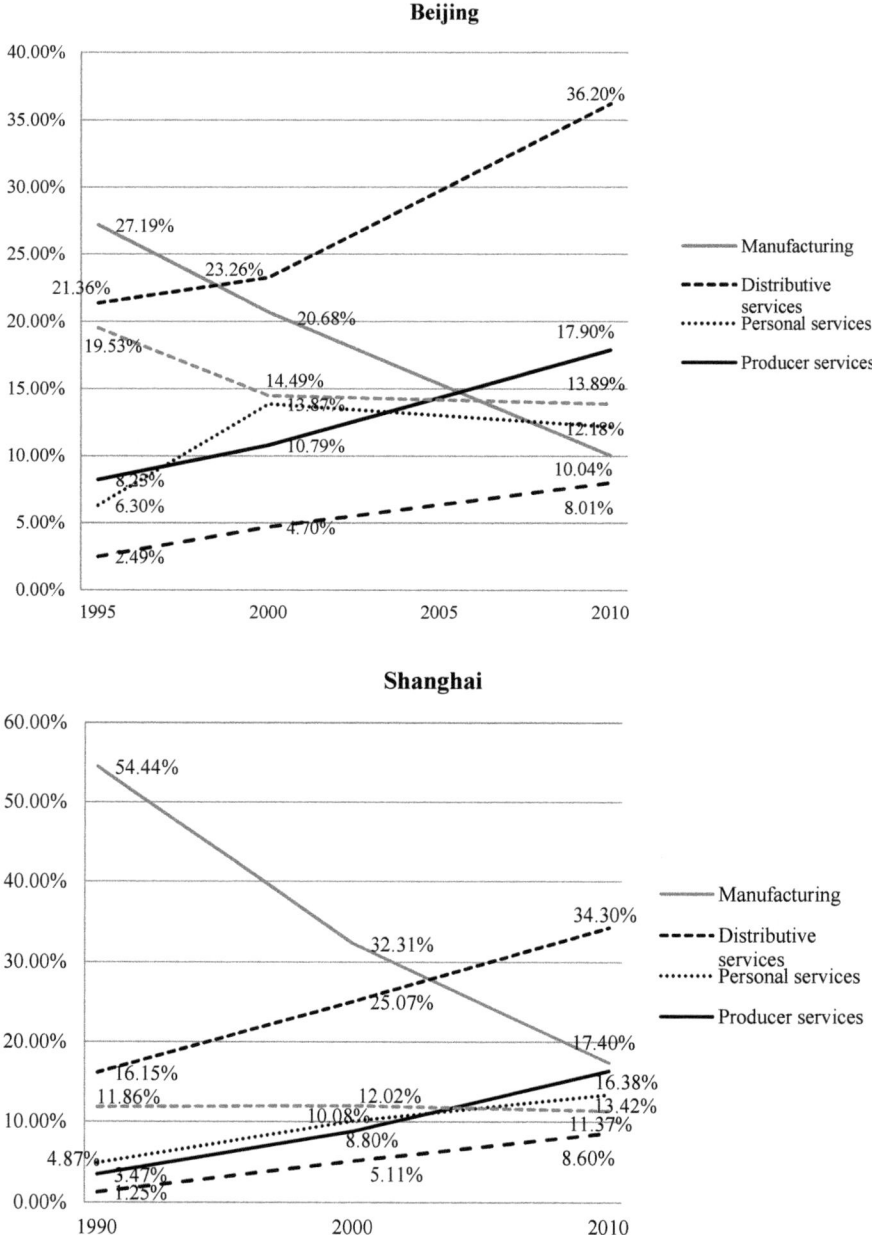

Fig. 2.1 Employment trend by sector in Beijing, Shanghai, Guangzhou, Chengdu 1990–2010. *Sources* Computed based on the population censuses

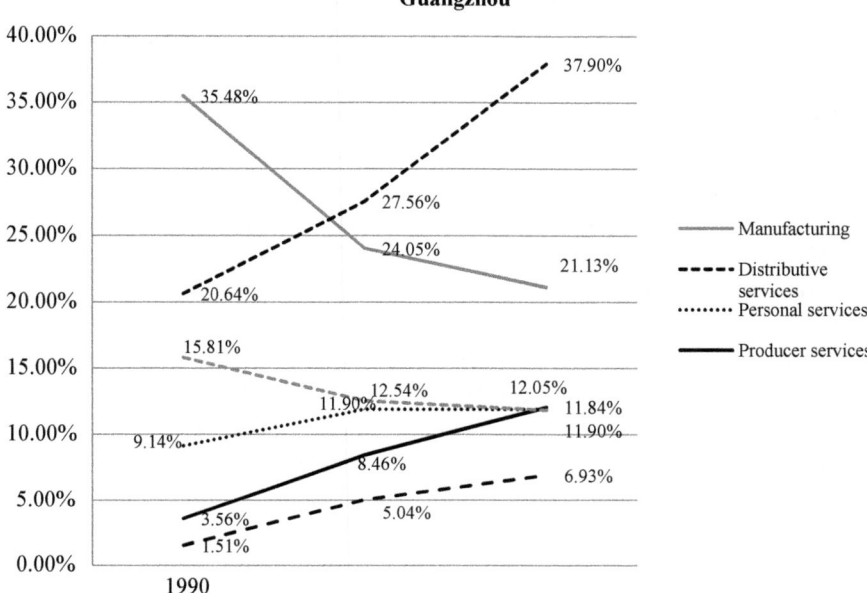

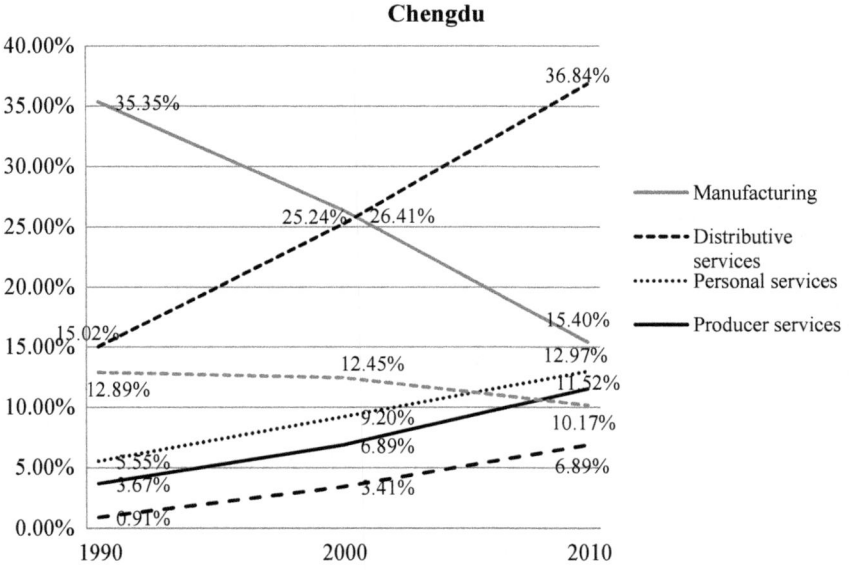

Fig. 2.1   (continued)

Table 2.1 Employment structures by sector in Chengdu: 1990, 2000 and 2010

	1990		2000		2010		Change 1990–2010	
	No	%	No	%	No	%	No	%
Farming, forestry, animal husbandry and fishery	210,235	15.86	130,820	8.15	19,080	0.76	−191,155	−90.92
Mining and quarrying	4939	0.37	2600	0.16	8280	0.33	3341	67.65
Manufacturing	468,594	35.35	424,120	26.41	384,360	15.4	−84,234	−17.98
Utilities	34,103	2.57	64,580	4.02	41,090	1.65	6987	20.49
Construction	107,853	8.14	104,620	6.52	248,560	9.96	140,707	130.46
Distributive services	199,141	15.02	405,290	25.24	919,460	36.84	720,319	361.71
Transport, storage, and postal services; information and communication	62,885	4.74	103,720	6.46	245,040	9.82	182,155	289.66
Wholesale and retail	136,256	10.28	301,570	18.78	674,420	27.02	538,164	394.97
Personal services	73,522	5.55	147,800	9.2	323,780	12.97	250,258	340.39
Accommodation and catering	21,832	1.65	83,920	5.23	162,980	6.53	141,148	646.52
Culture, sports, and entertainment	12,113	0.91	21,670	1.35	35,750	1.43	23,637	195.14
Residential, repair and other services	39,577	2.99	42,210	2.63	125,050	5.01	85,473	215.97
Producer services	48,697	3.67	110,570	6.89	287,530	11.52	238,833	490.45
Finance and insurance	9607	0.72	33,550	2.09	84,500	3.39	74,893	779.57
Real estate	2423	0.18	21,270	1.32	87,360	3.5	84,937	3505.45
Business and professional services	36,667	2.77	55,750	3.47	115,670	4.63	79,003	215.46
Social services	170,871	12.89	199,900	12.45	253,880	10.17	83,009	48.58
Education	75,895	5.73	79,860	4.97	106,680	4.27	30,785	40.56
Health care and social insurance	35,103	2.65	47,070	2.93	59,710	2.39	24,607	70.1

(continued)

Table 2.1 (continued)

	1990		2000		2010		Change 1990–2010	
	No	%	No	%	No	%	No	%
Public administration and social organizations	59,873	4.52	72,970	4.54	87,490	3.51	27,617	46.13
Other services	5535	0.42	3950	0.25	9690	0.39	4155	75.07
All others	2010	0.15	8940	0.56	0	0	−2010	−100
Total employed	1,325,500	100	1,605,790	100	2,495,710	100	1,170,210	88.28
Population aged 15 and over	1,855,747		2,782,110		4,543,200		2,687,453	

Note The census data for the industrial population in 1990 were based on a total population enumeration, whereas the 2000 and 2010 population were estimated by multiplying the census data by ten. *Sources* Computed based on the population censuses

greater autonomy in both labor recruitment and wage setting. Increasingly, system-generated privilege has been dismantled, and potential employees must compete flexibly [18, 26, 30, 31, 34]. The reforms resulted in extensive layoffs in departments and enterprises that were subject to restructuring and increased the unemployment risk in the city. Since the 1990s, particularly after the labor law was passed in 1994, the unemployment rate has been increasing annually. Based on the population accepting the Subsidized Minimum Living Standard, at least 22 million households, or 6% of urban citizens, may have suffered from underemployment in 2004 [2].

The two decades under study witnessed roughly a reverse between the development of the primary and secondary versus the tertiary and quaternary sectors in the city. Whereas the labor force in the four service categories expanded from 37.14% in 1990 to 71.51% of total employment in 2010, the number of those employed in primary and secondary industries shrank from 62.3 to 28.1%. This change indicates that within these two decades, an additional 1,292,419 people either transferred out of the manufacturing sector or newly arrived in the city and joined various services. This number is much larger than the 275,389 workers who left their manufacturing and agricultural jobs. This asymmetric change implies a tremendous influx of labor into the city as a consequence of urbanization. In 2010, service jobs accounted for more than 1.7 million of the city's employed population of 2.5 million.

Producer service workers in Chengdu in 2010 represented 4.9 times the number in 1990, which is faster growth than the other three categories of service industry in the city. This multiple is equivalent to an added labor force of 238,833 persons. Within producer services, real estate management has developed the most rapidly, increasing by more than 35 times in the number of jobs since 1990, followed by the still-remarkable growth in the financial and insurance industries (7.8 times) (Table 2.1). However, in 2010, these two sectors represented only 6.89% of the entire labor force in Chengdu (approximately 170,000 employees), compared with 8.6% (279,220 employees) in Shanghai (see Fig. 2.1). Although lower in scale, producer services in Chengdu grew faster (4.9 times) than those in the same sector in Shanghai (2.7 times). Within the service industries, personal services have rapidly expanded, increasing by 3.4 times from 1991 to 2010. However, the proportion of public service workers consistently decreased, from 12.89% of the entire employed population in 1990 to 10.17% in 2010 (Fig. 2.1). This result is consistent with [11] findings for Shanghai, which attributed the result to the economic reform that downsized public sectors as well as state- and collective-owned enterprises during the 2000s. The labor force generated by distributive services represented the largest portion of local employment in Chengdu (36.84%), the majority of which were in the wholesale and retail industries.

2.1.2 Changing Occupational Structure

The occupational structure depicts further particularities in the socioeconomic classification in China, derived from the status of not only the industrial transformation but also the reform of the economic system. The share remains far more consistent of professionals and managers in urban society in the four cities. Specifically, in Shanghai, when the share of low-paid production workers shrank from 50.24% of total employment in 1990 to 14.46% in 2010, the ratio of managers and professionals increased from only 22.91 to 31.01% (Fig. 2.2). In comparison, the share of production workers in Chengdu declined by only 17% (from 37.39% in 1990 to 20.81% to 2010), whereas the proportion of managers and professionals remained approximately 22% throughout the two decades. The ratio of managers and professionals even fell in Beijing (from 36.91 to 28.07%) and Guangzhou (from 24.07 to 22.86%). In contrast, the cities all witnessed notable growth in the proportion of low-paid sales and service workers, among which Chengdu had the highest rate (from 16.95% in 1990 to 42.01% in 2010).

Table 2.2 shows that in absolute numbers, managers and professionals in 2010 expanded by 77.39% and 93.47%, respectively, of the 1990 population in Chengdu, equal to 42,348 and 227,933 people, respectively. By 2010, 568,870 of the 2.5 million workers in the main urban districts of Chengdu were working in the two most highly ranked occupations. Nevertheless, among low-paid workers, only agricultural workers experienced a reduction in numbers throughout the twenty years from 1990 to 2010 across all urban districts of Chengdu. Although the production worker share among all occupations was decreasing, its population was still greater in 2010 than it was in 1990. The number of sales and service workers more than tripled from 1990 to 2010, in line with the change in the number of clerical workers, which was more than twice the number it had been in 1990. In 2010, low-paid service workers accounted for the largest labor pool in the city, totaling over 1 million. The rapid growth in low-skilled sales and service workers may have been partially derived from the increase in white-collar workers, who have generated demand for personal services; a similar relationship has occurred in advanced postindustrial cities. However, in Chengdu, the

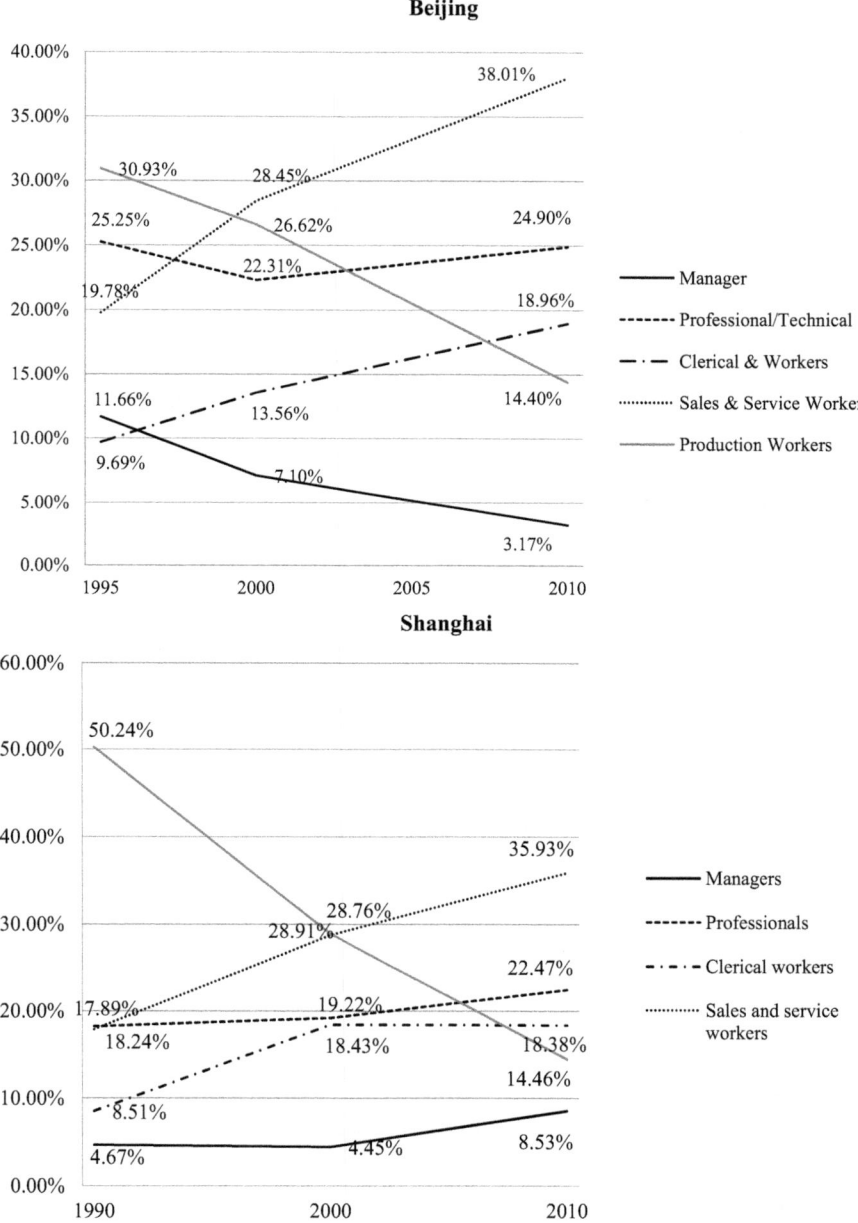

Fig. 2.2 Employment trends by occupation in Beijing, Shanghai, Guangzhou, Chengdu 1990–2010. *Sources* Computed based on the population censuses 1990, 2000, 2010

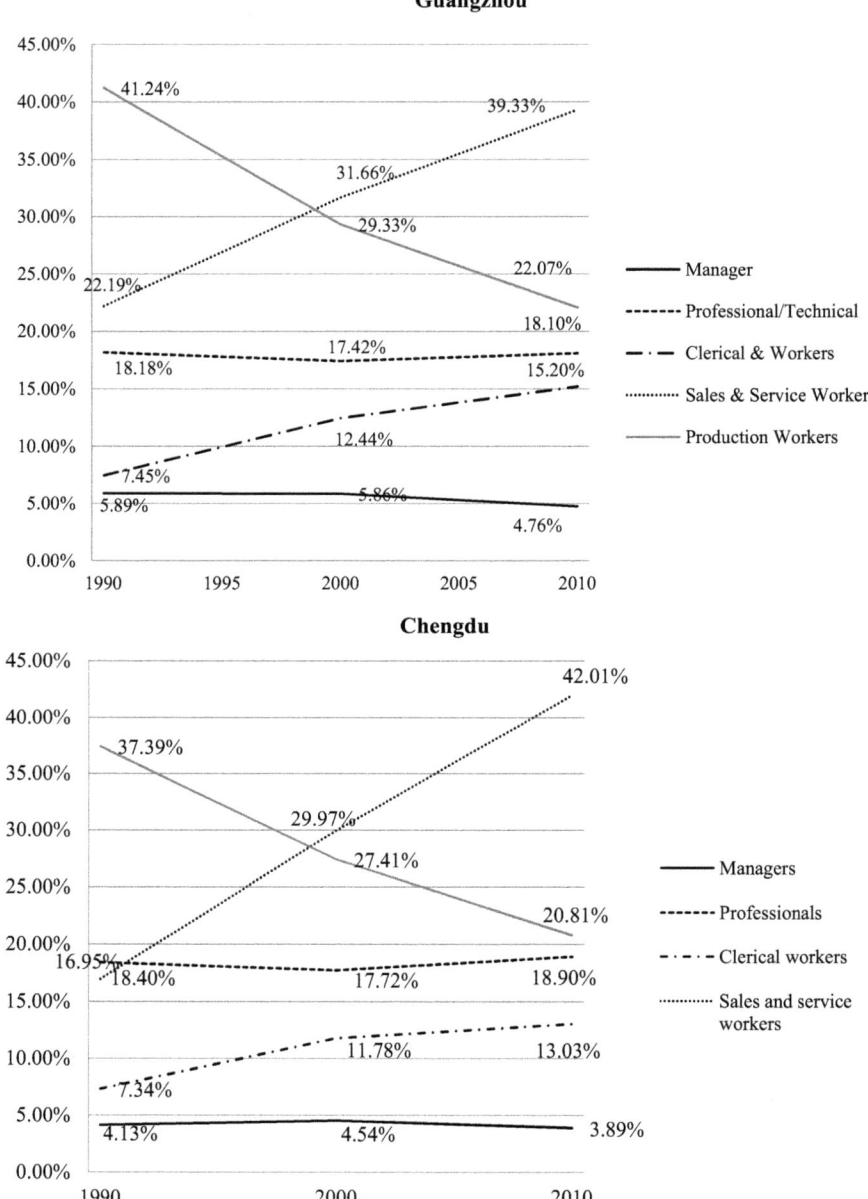

Fig. 2.2 (continued)

Table 2.2 Change in the employment structure of Chengdu by occupation 1990–2010

	1990		2000		2010		Change 1990–2010	
	No	%	No	%	No	%	No	%
Administrators and managers	54,722	4.13	72,950	4.54	97,070	3.89	42,348	77.39
Professionals and technicians	243,867	18.4	284,500	17.72	471,800	18.9	227,933	93.47
Clerical workers	97,336	7.34	189,120	11.78	325,230	13.03	227,894	234.13
Sales and service workers	224,617	16.95	481,240	29.97	1,048,360	42.01	823,743	366.73
Agricultural workers	206,353	15.57	130,540	8.13	16,670	0.67	−189,683	−91.92
Production workers	495,647	37.39	440,070	27.41	519,240	20.81	23,593	4.76
Others	–	–	7370	0.46	17,340	0.69	–	–
Total employed	1,322,542	100	1,605,790	100	2,495,710	100	1,173,168	88.71

Note The census data for the industrial population in 1990 were based on a total population enumeration, whereas the 2000 and 2010 population were estimated by multiplying the census data by ten.
Source Computed based on the population censuses

growth is more likely to be the result of economic strategies that encourage commercialization and consumerism and the influx of a large number of poorly educated migrants to the city from rural areas.

The evidence shows a lower growth rate for professionals and managers in Chinese cities from the 1990s to 2000s than was seen in advanced industrial economies in the 1970s. Ley [12] revealed a growth rate of 30% for professionals and 65% for managers in British Columbia within just five years, from 1971 to 1975. He [11] also showed that the increase in professionals and managers in terms of both number and ratio in Shanghai from 1990 to 2000 was much smaller than that for their Western counterparts in the 1970s and 1980s. Notwithstanding, in 1990, the percentage of professionals and managers in Chengdu (22%) already exceeded the level in the province of British Columbia in 1975 (20%).

This situation represents the influence of enterprise reform on socioeconomic restructuring during China's transitional period. Since the late 1980s, enterprise reform has encouraged the decentralization and marketization of enterprise operations. It has resulted in the privatization or dismantling of small-scale, state- or collective-owned firms and the development of private companies in both urban and rural areas [19, 20, 32]. Corporate restructuring has been accompanied by individual job transfers in former state sectors and labor force enlargement in private sectors. The number of employees in state-owned enterprises decreased by 49 million, from

113 million in 1995 to 64 million in 2004 ([2], p. 176). Fundamentally, the new labor market redistributes life opportunities among the employed population, whose positions and salaries have been conditioned by personal skills, educational levels, labor productivity and the economic performance of an enterprise [2, 26, 34]. In this context, high-ranking employees affiliated with the old system might initially have experienced profound career changes, for example, being either filtered out of or developed as members of the new sectors or benefiting from participating in the dual system. Turbulence in the job market, reflected in employment retrenchment and subsequent labor mobility, could increase or decrease the presence of highly educated workers in major urban districts.

The stage of industrial transformation, accompanied by the unique history of economic system reform, is important contextual information, as it could generate contradictory socioeconomic dynamics in the manifestations of gentrification. Undoubtedly, large Chinese cities have grown overwhelmingly in producer service industries over the past two decades. Meanwhile, manufacturing closures have laid the foundation for the working-class retreat from the inner city. However, high-skilled services expanded simultaneously with low-skilled services. The turning point of a typical postindustrial society will take time to achieve. Compared with global cities in the West in the 1970s, knowledge-based service industries in cities were in a less dominant position in China by 2010. Socioeconomic realities may challenge the specific segment of high-end service classes that comprise the majority of gentrifiers in China.

Economic reform further complicates the mode of socioeconomic restructuring. Various socioeconomic groups have been experiencing a process of restratification along with a transition from a planned production to a market-oriented system and the subsequent reconfiguration of job opportunities among individuals. The socioeconomic restratification in progress may generate inquiries into the social composition of middle-class gentrifiers and the source of their collective motives in inner-city resettlement. These questions lead this study to stress the second context of gentrification with regard to cultural transformation and middle-class formation in a post-socialist society.

2.2 Individual Consumption and Middle-Class Formation in a Transitional Society

The cultural tide of postmodernism is another critical context for gentrification in postindustrial cities. The classical gentrification model is bred in a liberal ideology of the cultural innovation of urban lifestyles [12–14]. Caulfield [3], who questioned the use of the term gentrification, deemed middle-class resettlement in the older inner city a critical social practice of middle-class agents. This resettlement contributes to the creation of an emancipatory city and aims to compete with the dominant lifestyle characterized by a standardized corporate form and suburban lifestyle. Ley

[13] analyzed urban pioneers and the new middle class's engagement with historic preservation, neighborhood development and progressive political reform from the 1960s to the 1970s, contrasting these trends with the economistic values and mass culture that prevailed under Fordist mass production. Zukin [36] scrutinized the emergence of loft-living in Lower Manhattan in New York City, with an intricate thesis about the relationship between cultural production initiated by artists at the outset, followed by capital accumulation through recreating built environments by subsequent developers and the state in a period of deindustrialization.

2.2.1 "Classless" Society

Unlike postindustrial contexts, in post-socialist China, the relationship between housing consumption and class formation and division is pertinent to the sociocultural characteristics of socialist society and the composition of the new rich, which has been impacted by economic reform. First, under Mao's China, a lifestyle characterized by communal consumption and the collective living of cadres and workers in *danwei* compounds established an egalitarian base upon which social divisions were rarely embodied in habitus distinctions. Zhang [35] explained that the so-called egalitarian society was manifested precisely in homogenous living patterns, in low salaries and in an underdeveloped consumer society in socialist China. Consequently, the socialist society in China presented no clear classification of lifestyles and political attitudes, although it did have an explicitly hierarchical system of occupations. Within this historical social context, contemporary class society in China thus features the creation and circulation of cultural distinctions among social groupings.

Second, the pathway to economic reform has generated social differences among the new middle class in Chinese cities. The result can be traced back to Deng Xiaoping's middle-class politics, which tried to facilitate reform; this area has been largely underdeveloped by urban researchers compared to its economic counterpart. At the beginning of the reform, a basic national policy was proposed, which has to date served as a constant principle guiding China's social development objectives: to establish a moderately prosperous (*Xiaokang*, 小康) society. Together with two other important public policies at the central level, rural and urban household registration and the one-child policy, this basic national policy has deeply impacted the trajectory of social change in China. Deng's description of the "moderately prosperous society" is in contrast with the affluent society of developed countries and Mao's common prosperity (*Datong*, 大同). Deng was concerned that Chinese society, within the medium term (say by 2050), could by no means achieve the quality of life seen in advanced societies, but it could aim to be a middle-class society [6]. This policy is thus a moderate revision of the bold "four modernization" objectives offered by Mao (see also [15, 16, 27]). While Mao's *Datong* society reflected his idealist imaginings for a communist society, Deng was absolutely realistic in recognizing the impossibility of achieving the objective of common prosperity based on the economic level of pre-reform China. Another difference between *Datong* and *Xiaokang* is that while

Datong society advocates for an absolutely public notion of ownership and distribution, *Xiaokang* society recognizes the necessity of individual wealth and private
assets and stresses family values. According to [16], Deng considered these concepts
to be much more acceptable to the masses and thought that they could be used to
deliver market principles to the masses in accordance with economic reform.

2.2.2 Emerging New Rich

Based on this background, the slogan "get rich first and achieve common prosperity later" was promoted by Deng in the 1980s as a realistic short-term path
toward an economic transition and *Xiaokang* society, with the ultimate goal of
achieving common prosperity [33]. Significantly, the strategy promotes the formation of specific groups in particular sectors and regions that would benefit from
reformist policies. These favored groups are the so-called "new rich" and "the newly
emerging middle class" in China [33]. Tomba [27] showed that, in the second half
of the 1990s, salaries for professionals in the public sector rose by 168%, which was
40% higher than the average. Tong [28] revealed that average household incomes in
Shanghai, Guangzhou and Beijing more than doubled in the two years from 1993 to
1995, whereas in the late 1990s, up to 0.3 million workers were expected to become
redundant in one year in Shanghai (p. 33).

Based on the dataset from the *Statistical Yearbook*s in Beijing, Shanghai,
Guangzhou and Chengdu, Fig. 2.3 shows the changes in the disposable income
of urban households. The figure divides the income groups into five quintiles, with
additional information on the top and bottom 10% in certain years for Chengdu
and Shanghai. Since approximately 2002, the households in the five quintiles in the
four cities all experienced faster income increases than they saw in the 1990s. As a
household earned more wealth, its income rose. As a result, a clear polarization has
emerged between the high-income other groups, which has been sustained since the
early 2000s. The tendency is less obvious in Chengdu and Guangzhou than in it is
Shanghai and Beijing.

The egalitarianism of socialist society has disappeared; China's economic reform
has generated some more affluent groups in urban society who can act as the principal
force in the consumer market. These groups favored by reformist policies are the
so-called new rich and constitute the emerging middle class in China [8, 15].

Notably, this economic reform generated various sources, formal and informal,
of affluence. The pre-reform elites can be separated by the direction of their social
mobility within the processes of job change, such as either consolidating an elite
social status or disappearing in competition. Walder [29] suggested that the opportunity structure for pre-reform elites varies in transitional economies and is determined
by, first, the divergent trajectories of political reform and, second, by the systematic
reform of the public ownership of assets. The peculiarity of reform in China lies
in that hitherto, the Chinese authority showed no interest in political reform and
maintained mixed ownership types and production relations. Different economic

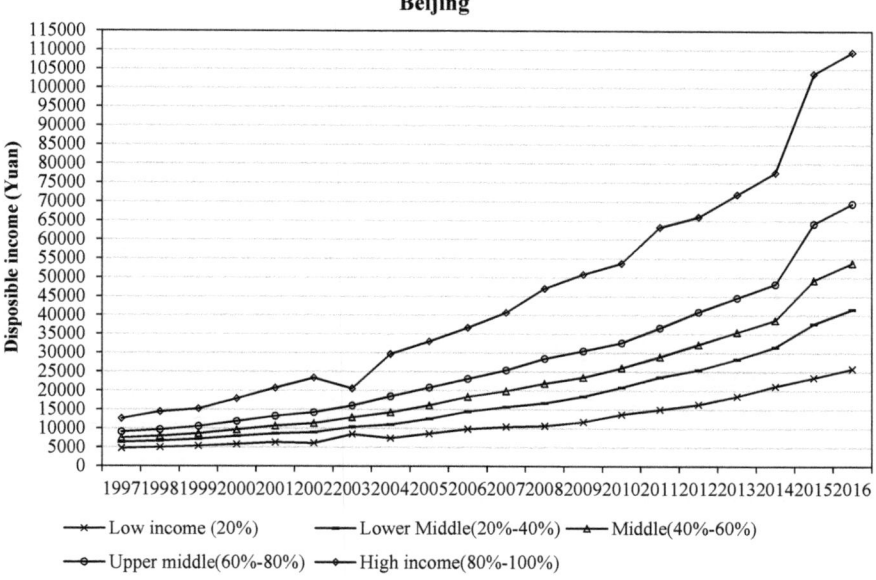

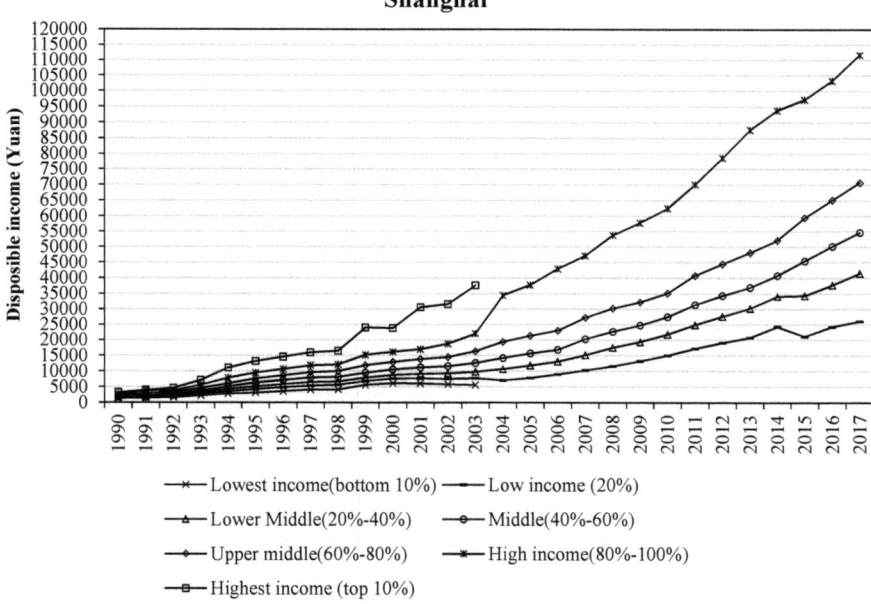

Fig. 2.3 Average disposable income per capita in Beijing, Shanghai, Guangzhou, and Chengdu 1990–2018. *Note* The information for the top and bottom 10% disposable income per capita is missing for Beijing and Guangzhou. Since 2014, the *Chengdu Statistical Yearbooks* have no longer classified income groups in Chengdu. *Source* Computed based on the *City Statistical Yearbooks* 1998–2018

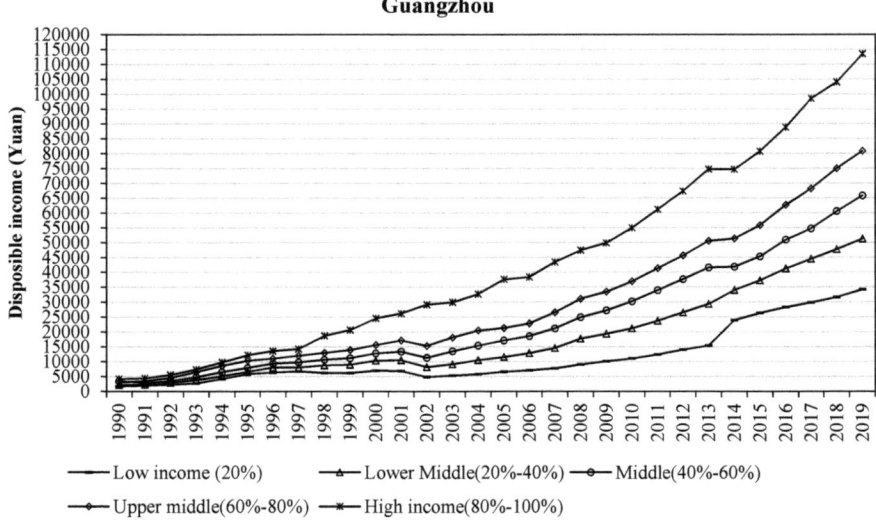

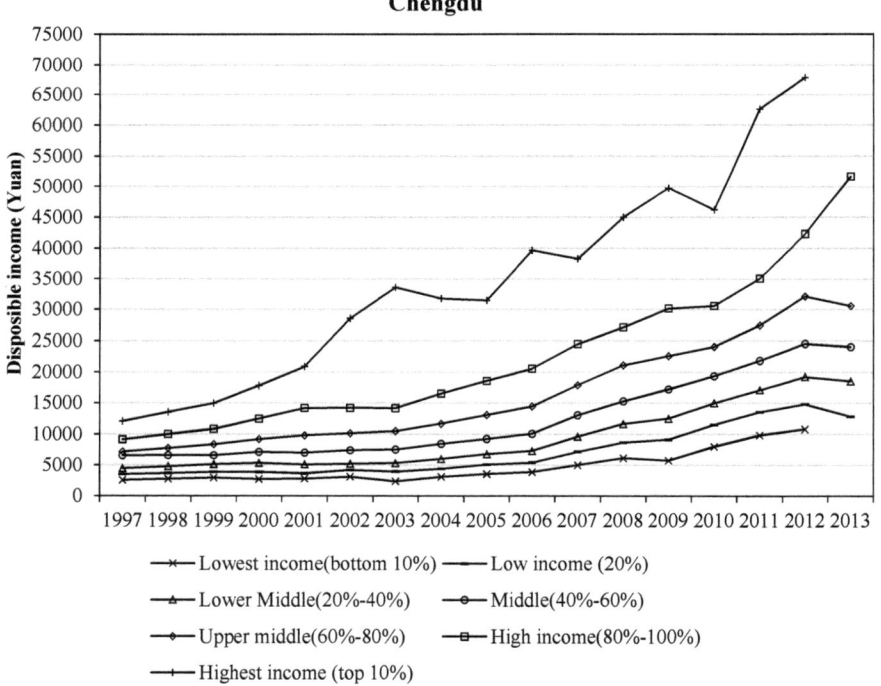

Fig. 2.3 (continued)

systems and workplaces impact the path of success, political ideology, wealth accumulation and lifestyle ([17], p. 265), which creates social differences within the new middle class. The most salient case is that the first generation of the new rich in China is not composed of well-educated and skilled professionals or managers but of petty businessmen and private entrepreneurs who have relatively low educational attainments [17]. Nevertheless, a *danwei* with stronger political and social ties can create better life opportunities for its employees. Managers in large (vs. small) *danwei* can possess substantially more wealth and social prestige because of the differential power they have in acquiring resources. The "cadre-entrepreneur," who occupies a high-end position in both private firms and public administrations, is a particular social construction of the transitional economy ([21], p. 269). As a result, the new middle class in China lacks internal recognition, in particular, recognition embodied in the uneven distribution of educational attainment and cultural character (or symbolic capital).

2.2.3 Consumer Revolution of the New Rich

Economic liberation has accompanied the post-reform social policies of the Chinese Communist Party, which has transitioned from representing the release of productive forces to the liberation of consumption forces [4, 27]. The concept of a moderately prosperous society places family happiness, self-interest and material well-being at the core of individual achievement rather than public concerns, political participation or human development [1, 16]. Lu argued that, in a society in which policies create enriched opportunities and diversified channels for pursuing affluence, with "everyone caring about themselves, people want more than what and how they can achieve it, while not [being concerned with] what the others get and if it is fair" ([16], p. 114). Under this political ideology of social development, a "consumer revolution" has swept Mainland China and, in particular, mobilized individual consumption among China's new rich ([5], p. 692).

Figure 2.4 presents the changing consumption structure of the urban population in the four large cities since 1990. Over the years, personal daily consumption increased rapidly. For example, in Shanghai, the average consumption expenditure of urban households was 1937 yuan per capita in 1990 in Shanghai (Chengdu 1680.77) and 23,200 yuan per capita in 2010 (Beijing 19,934, Guangzhou 25,011.61, and Chengdu 15,510.91), compared with an average income of 2050.2 yuan (Guangzhou 2593.08) in 1990 and 31,838 yuan per capita in 2010 (Beijing 29,073, Guangzhou 30,658.49, and Chengdu 20,835.34). Expenditures reached 46,015 yuan in 2018 (Beijing 42,926, Guangzhou 42,180.96, and Chengdu 27,312.05), compared with an average income of 64,183 yuan (Beijing 67,990, Guangzhou 59,982.1, and Chengdu 42,127.85). The consumption/income ratio declined in all cities, which implies, on the one hand, a large consumption potential for residents in these cities and, on the other hand, an even faster increase in investment expenditures, such as housing.

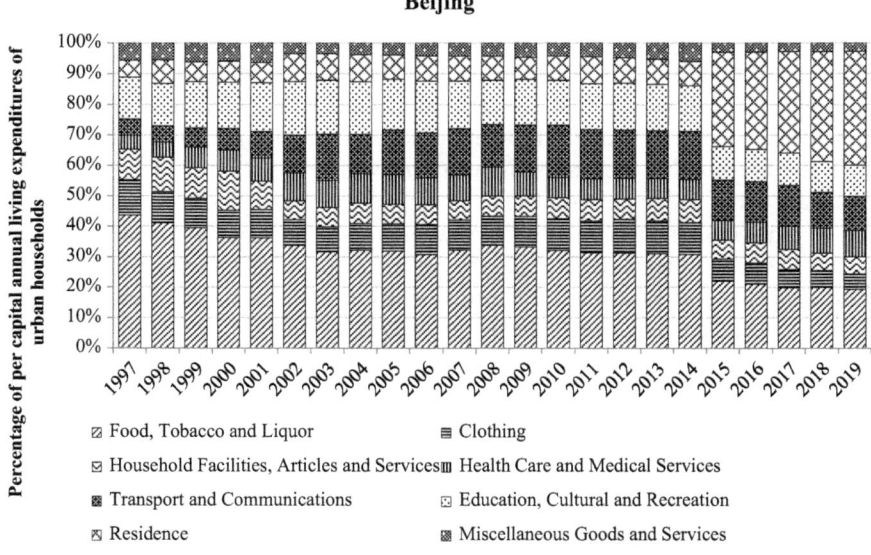

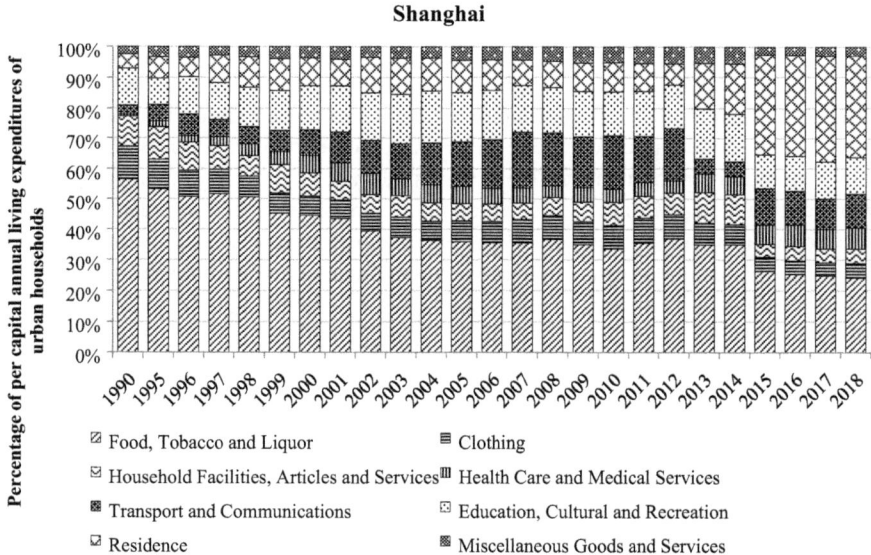

Fig. 2.4 Changing consumption pattern of urban households in Beijing, Shanghai, Guangzhou, and Chengdu 1990–2010. *Source* Computed based on the *City Statistical Yearbooks* 1998–2018

Guangzhou

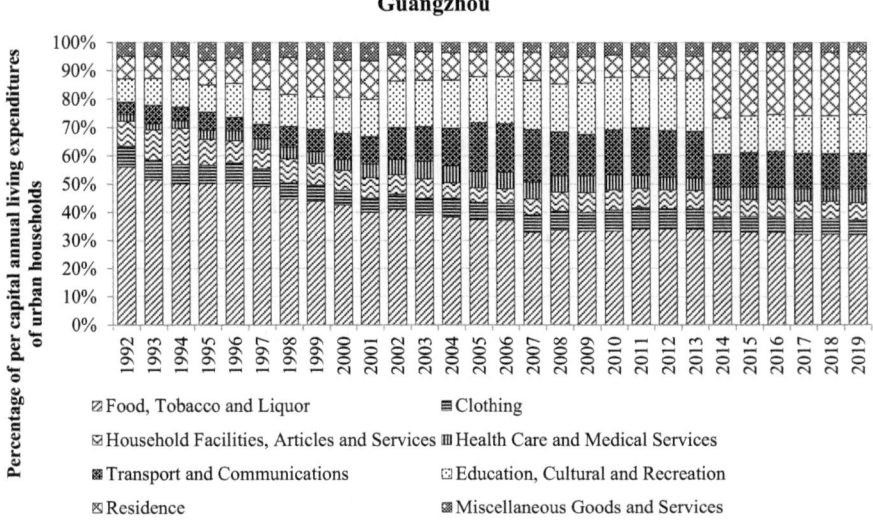

Chengdu

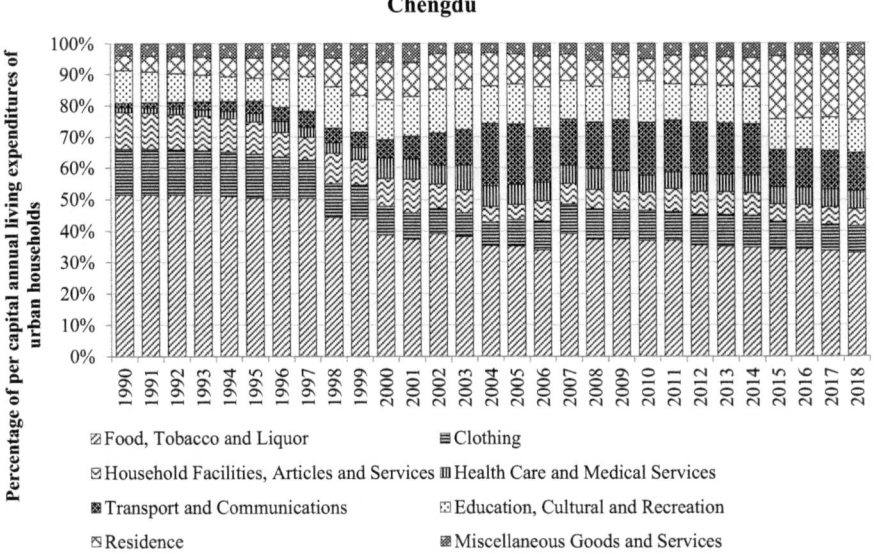

Fig. 2.4 (continued)

Specifically, these 20–30 years witnessed an obvious reversal between the propor-tions of consumption expended for basic demands (food, clothing and house-hold facilities) and for residences (excluding the cost of housing purchases and construction), education/culture/recreation, transportation and health care. As the new statistical yearbooks classified housing rent into residence consumption, resi-dent consumption grew dramatically in all of these cities in 2015. In Shanghai, the average spending on residence, education, transportation, and health care accounted for only 20.13% of personal consumption expenditure in 1990, while this segment expanded to 63.45% in 2018 (Beijing: 28.97% in 1997 to 67.25% in 2019). The change in Guangzhou and Shenzhen was relatively moderate (Guangzhou: 31.39% in 1997 to 53.33% in 2018; Chengdu: 18.18% in 1997 to 49.05% in 2018). When the changing statistical classification is ignored, the fastest expansion for the four sections was mostly in the 2000s, responding to the radical development of Chinese cities in this period. During this period, the fastest expansion was the cost of transportation and communication, followed by the cost of education, culture and recreation. The change in consumption mode shows the changing lifestyle of urbanites, with increasing attention given to cultural, spiritual and personal development demands.

Thus, post-socialist societal transition in Chinese cities has animated a cohort of high-income consumers who have been exposed to class (re)making, particularly cultural redefinition and restratification. Conspicuous consumption practices, rather than the latent means of production, have become empirically necessary and viable for a class analysis in China. The changing landscapes of newly built communities have become a decisive, practical field in which the new rich proclaim their cultural distinction and form new patterns of collective interests; mobilization; and, consequentially, collective conflicts [22, 23, 27, 35].

Gentrification in China is thus occurring in a society wherein so-called middle-class distinctions are emerging and forming. Classical pioneer-led gentrification in Western cities conveys a cultural claim by a new middle class standing in opposition to suburbanites. In contrast, state-facilitated gentrification in China, through the massive demolition and reconstruction of urban places and the accompanying social reorganization in place, can be the very force driving class formation. This phenomenon suggests that, to explain gentrification in China, one should not consider the process as simply an interaction between two *established* classes. Instead, it represents a highly dynamic social change that reflects the relationship between inner-city urbanism, class formation, replacement and displacement.

References

1. Anagnost A (2008) From "class" to "social strata": grasping the social totality in reform era China. Third World Q 29(3):497–519
2. Cai F, Park A, Zhao Y (2008) The Chinese labor market in the reform era. In: Brandt L, Rawski T (eds) China's great economic transformation. Cambridge University Press, Cambridge, pp 167–214
3. Caulfield J (1994) City form and everyday life: Toronto's gentrification and critical social practice. University of Toronto Press, Toronto
4. Davis D (2000) The consumer revolution in urban China. University of California Press, Berkeley
5. Davis D (2005) Urban consumer culture. China Q 183:692–709
6. Deng X (1983) Deng Xiaoping Wenxuan, Dierjuan (Selected works of Xiaoping Deng), vol 2. People's Publishing Housing, Beijing

7. Edgington DW, Haga H (1998) Japanese service sector multinationals and the hierarchy of Pacific Rim cities. Asia Pac Viewp 39:161–178
8. Goodman DSG (2014) Middle class China: dreams and aspirations. J China Polit Sci 19:49–67
9. Hamnett C (1986) The changing socio-economic structure of London and the South East. Reg Stud 20:391–406
10. Hamnett C (2003) Gentrification and the middle-class remaking of Inner London, 1961–2001. Urban Studies 40(12):2401–2426
11. He S (2010) New-build gentrification in central Shanghai: demographic changes and socioeconomic implications. Popul Space Place 16(5):345–361
12. Ley D (1980) Liberal ideology and the postindustrial city. Ann Assoc Am Geogr 70(2):238–258
13. Ley D (1996) The new middle class and the remaking of the central city. Oxford University Press, Oxford
14. Ley D (2012) Social mixing and the historical geography of gentrification. In: Bridge G, Butler T, Lees L (eds) Mixed communities: gentrification by stealth? Policy Press, Bristol, UK; Chicago, IL, pp 53–68
15. Li JR (2003) Lun Quanmian Jianshe Xiaokang Shehui (On the development of Xiaokang society). China Soc Sci 1:4–12
16. Lu HL (2010) The Chinese middle class and xiaokang society. In: Li C (ed) China's Emerging middle class: beyond economic transformation. The Brookings Institution Press, Washington, DC, pp 104–133
17. Lu XY (2002) Zhongguo Shehui Jieceng Baogao (Social stratification in contemporary China). Social Science Academic Press, Beijing
18. Meng X (2000) Labour market reform in China. Cambridge University Press, Cambridge
19. Naughton B (1994) Chinese institutional innovation and privatization from below. Am Econ Rev 84(2):266–270
20. Nee V, Opper S (2012) Capitalism from below: markets and institutional change in China. Harvard University Press, Cambridge, MA
21. Nees V (1991) Social inequalities in reforming state socialism: between redistribution and markets in China. Am Sociol Rev 56(3):267–282
22. Pow CP (2009) Gated communities in China: class, privilege and the moral politics of the good life. Routledge, London
23. Pow CP, Kong L (2007) Marketing the Chinese dream home: gated communities and representations of the good life in (post-) socialist Shanghai. Urban Geogr 28(2):129–159
24. Sassen S (1991) The global city: New York, London, Tokyo. Princeton University Press, Princeton
25. Scott AJ (2011) Emerging cities of the third wave. City 15(3–4):289–321
26. Tomba L (2002) Paradoxes of labour reform: Chinese labour theory and practice from socialism to the market. University of Hawaii Press, Honolulu
27. Tomba L (2004) Creating an urban middle class: social engineering in Beijing. China J 51:1–26
28. Tong L (1998) Consumerism sweeps the mainland. Mark Manag 6(4):32–35
29. Walder AG (2003) Elite opportunity in transitional economies. Am Sociol Rev 68(6):899–916
30. White G (1987) The politics of economic reform in Chinese industry: the introduction of the labour contract system. China Q 111:365–389
31. White G (1988) State and market in China's labour reforms. J Dev Stud 24(4):180–202
32. White G (1993) Riding the tiger: the politics of economic reform in post-Mao China. Stanford University Press, Stanford, CA
33. Yan ZM, Liu J (2002) Deng Xiaoping Xianfu Gongfu Lilun yu Woguo Dangqian Xianfu Qunti Wenti (Deng Xiaoping's theory about first rich and common rich and problems of group of the first riches). Expand Horiz 2:4–6
34. Yueh L (2004) Wage reforms in China during the 1990s. Asian Econ J 18(2):149–164
35. Zhang L (2010) In search of paradise: middle-class living in a Chinese metropolis. Cornell University Press, New York
36. Zukin S (1982) Loft living: culture and capital in urban change. The Johns Hopkins University Press, Baltimore

Chapter 3
The Geography of Gentrification in Chengdu, 2000–2010

The choice of indices with which to measure gentrification has varied with the literature and has been impacted by the working definitions of gentrification employed by scholars. Analysts can select specific indices and measurement methods based on their theoretical perspective and research purpose [16]. For instance, Ley [10] defined the socioeconomic upgrading of neighborhoods as a decisive factor marking the existence of gentrification and used changes in occupational and educational status as two of its main measures (see also [5]). Following Smith's [15] production-side explanation of gentrification, authors have adopted indicators representing reinvestment as an effective measure of gentrification. Wyly and Hammel's [18] work, for example, was assisted by household-level mortgage data in New York City. Generally, changes in land and building values have been used to identify gentrification in a neighborhood [1, 4, 12]. Other studies have used mixed methods and indices to improve their accuracy in identifying gentrified neighborhoods. For instance, Walks and Maaranen [16] included indicators measuring both socioeconomic upgrading and capital investment. Lees [8] combined ethnographic fieldwork and statistics when generating evidence for supergentrification led by financiers in New York City. Wyly and Hammel's [17] method also drew on literature reviews, field research and a multivariate discriminant analysis of the socioeconomic characteristics of neighborhoods.

This study takes the position that the most solid conceptual foothold for gentrification is the upgrading of places in terms of social class. However, in practice, measuring social class in China remains problematic. Because the labor force allocation system has been fundamentally reformed into a competitive labor market, the unemployment rate throughout the city has continued to soar since the mid-1990s; meanwhile, the employed population has been restructured across new emerging sectors based on a market-oriented production system and among old sectors based on a planned system [3]. As a result, the percentage change in professionals and managers in large Chinese cities is much lower than that seen in advanced economies since the 1970s. Labor market reform may cause an underestimation of the degree

---

This chapter has been rewritten and published in Geoforum (2018, 93, pp. 120–132).

© Springer Nature Singapore Pte Ltd. 2022
Q. Yang, *Gentrification in Chinese Cities*, Urban Sustainability,
https://doi.org/10.1007/978-981-19-2286-2_3

of change in the socioeconomic population of a neighborhood. Moreover, internal social differentiation in the status of employees with similar occupations in companies or organizations in terms of—for example, educational levels, incomes, values, etc.—increases the uncertainty of defining the property and size of the middle class in China [6]. A direct impact of these problems is that researchers have deployed diversified indicators and have gained highly divergent estimations of the volume of the middle class in urban China. For example, a measurement according to indicators of consumption levels estimated the middle class as representing 54% of the overall urban population in 2012 [14]. Meanwhile, a prediction based on occupation and income estimated the middle class as constituting only 23% of the urban population in the mid-2000s [13]. When an indicator capturing self-identification is added, the number continues to decrease [11, 19].

In light of the essence of gentrification, one main purpose of the measurement in this study is to compare the degree of social structural upgrading in place. Unlike the above literature, this work does not aim to accurately estimate the size of the Chinese middle class. In effect, this study confirms that the attributes of the so-called middle class are essentially unreliable at this stage, as it is subject to varied constructions and self-ascription. Furthermore, the concerns mentioned above do not necessarily reject the use of the ratio of the change in professionals and managers as an important index of neighborhood social upgrade.[1] Thus, following Ley [10], this study continues to adopt the two main indices of change in social class: the change in the location quotient for local residents over 6 years of age with university degrees and that for professionals and managers in the subdistrict.[2] However, by creating a gentrification index based on these two indices, this study offers a relatively conservative estimate of the volume of gentrifiers.

Compared with the relative percentage change, a change in the location quotient (LQ) can be more effective for measuring gentrification in China (see also [7]). The location quotient index expresses the comparative advantage of a place in containing a population relative to the average level of a larger territory. The index of the changing location quotient thus essentially reflects the difference in the concentration of a social class in a place throughout a period, while the relative percentage change restates that change against the base at the start of the period for that location. The former can avoid the information bias caused by rapid demographic densification and turbulence in the labor market throughout the entire city. With census data for 2000 and 2010, this study created location quotients for the population with a university degree and for professionals and managers in each *subdistrict* of the inner city, which is smaller than an urban district and larger than a neighborhood. All of the location quotients reflecting changes in the inner-city subdistricts are compared with changes in the five *main urban districts* of Chengdu established since 1990. The location quotients for

---

[1] In effect, income can be an optimum index of social upgrading in China. Unfortunately, the census data do not offer any information about personal or household incomes. The other social survey datasets that include income do not release household geographical information.

[2] The age group (i.e., residents over 6 years old) to which the ratio of the university population is measured follows the standard of the Population Census of China.

the two populations are calculated by dividing the percentages for a subdistrict by the percentages for the main urban districts. The change in the location quotient is estimated by directly subtracting the value in 2000 from that in 2010. A gentrification index is then created based on the arithmetic mean of the two change indices for the location quotients from 2000 to 2010.

Based on the quantitative results of this gentrification index, I then verified the gentrified locations through field observations of the visible landscapes of renovation and reinvestment. Moreover, using the yearbooks that documented the subdistrict changes, this study excludes factors that could have caused bias in the quantitative data. First, the study adjusted the educated population when the outmigration of universities and colleges in the subdistricts caused a sudden decrease in the population with a university degree during the census interval (Jianshelu, Wangjianglu).[3] The portion of change caused by campus outmigration was omitted from the gentrification index. Then, I focused on subdistricts in which most of the land parcels have been subject to commercial use (Yanshikou) and upgrading was due to a large proportion of new greenfield construction in the inner city (Shuangnan). These areas may necessitate data adjustments for certain variables in the two cases that are anticipated to have a significant influence on the correlations analyzed in the next section. Finally, concerning the gentrified neighborhoods identified in this study, I confirmed the reasons for the upgrades from the archives to avoid being misled by the above factors.

3.1 Gentrification Extent

Table 3.1 lists the population with a university degree and professionals and managers in 31 subdistricts in the inner city of Chengdu. In the last column, the composite change in the LQs indicates the magnitude of gentrification in these subdistricts in descending order. Core-gentrified locations in this study are identified as those in which the increase in the middle-class index is greater than inner-city upgrading (−0.01) and city-level change (0). Moreover, the LQ of core-gentrified locations must exceed the LQ change for Jinjiang District (0.24), which represents the highest level of socio-spatial upgrading in the five urban districts. A less-gentrified place is defined by a gentrification index higher than zero and lower than the index of Jinjiang District (0.24), while places with a gentrification index below zero are ungentrified.

During the 2000s, the inner city of Chengdu retained its prominent role in housing the middle class. In both 2000 and 2010, the proportion of the population with a university degree in the inner city was higher than it was for the entire urban district (LQ = 1.06 in 2000 and 1.22 in 2010). Comparable results are found for

---

[3] Based on the *Annual Report of Chengdu*, I first noted the number of university students in a subdistrict in 2000 and the number of *graduate* students. I then subtracted the two numbers from the total population and the population with a university degree in this subdistrict. Finally, the two estimated values were used to calculate the ratio of the population with a university degree in the subdistrict in 2000. This ratio was compared with the 2010 ratio in this subdistrict.

Table 3.1 Socioeconomic profiles of inner-city subdistricts in Chengdu

| | | Population over 6 and with university degree | | | | | Administrators/managers and professionals/technicians | | | | | Composite change 2000–2010 | |
| | | 2000 | | 2010 | | Change in LQ | 2000 | | 2010 | | Change in LQ | Absolute % | LQ |
		%	LQ	%	LQ		%	LQ	%	LQ			
Core gentrified	Lianxin	4.08	0.44	28.06	1.85	1.41	22.57	1.01	38.49	1.69	0.67	19.95	1.04
	Longzhou	4.08	0.44	22.01	1.45	1.01	22.57	1.01	36.40	1.60	0.58	15.88	0.80
	Niushikou	4.51	0.48	17.27	1.14	0.65	19.60	0.88	29.23	1.28	0.40	11.20	0.53
	Xinhuaxi	8.00	0.86	24.16	1.59	0.73	35.22	1.58	39.36	1.73	0.14	10.15	0.44
	Hejiangting	6.77	0.73	24.32	1.60	0.87	31.21	1.40	31.65	1.39	-0.01	9.00	0.43
	Fuqing	5.51	0.59	13.45	0.89	0.29	22.77	1.02	32.26	1.42	0.39	8.72	0.34
	Caoshijie	8.83	0.95	24.99	1.65	0.70	30.00	1.35	29.59	1.30	-0.05	7.88	0.32
	Wangjiaguai	11.85	1.27	30.01	1.98	0.70	36.36	1.63	35.94	1.58	-0.06	8.87	0.32
Less gentrified	Shuangnan	8.35	0.90	20.88	1.37	0.48	28.72	1.29	26.16	1.15	-0.14	4.99	0.17
	Shuijinfang	7.92	0.85	20.03	1.32	0.47	31.62	1.42	28.69	1.26	-0.16	4.59	0.15
	Jianshe	5.15	0.55	13.45	0.89	0.33	38.28	1.72	35.15	1.54	-0.18	2.59	0.08
	Xinhong	6.10	0.66	9.56	0.63	-0.03	27.82	1.25	32.11	1.41	0.16	3.88	0.07
Ungentrified	Shaocheng	11.91	1.28	21.90	1.44	0.16	36.42	1.64	33.28	1.46	-0.18	3.43	-0.01
	Jiangxijie	21.53	2.31	38.92	2.56	0.25	39.39	1.77	34.24	1.50	-0.27	6.12	-0.01
	Renminbei	8.40	0.90	16.69	1.10	0.20	32.18	1.45	27.80	1.22	-0.23	1.96	-0.01
	Shuangqiaozi	5.19	0.56	10.08	0.66	0.11	26.84	1.21	24.18	1.06	-0.14	1.12	-0.02
	Shuyuanjie	8.33	0.89	17.53	1.15	0.26	35.45	1.59	28.46	1.25	-0.34	1.11	-0.04
	Caotang	12.34	1.33	20.49	1.35	0.02	38.04	1.71	34.11	1.50	-0.21	2.11	-0.09
	Xiyuhejie	10.81	1.16	21.11	1.39	0.23	34.88	1.57	26.10	1.15	-0.42	0.76	-0.10

(continued)

Table 3.1 (continued)

	Population over 6 and with university degree					Administrators/managers and professionals/technicians					Composite change 2000–2010	
	2000		2010		Change in LQ	2000		2010		Change in LQ	Absolute %	LQ
	%	LQ	%	LQ		%	LQ	%	LQ			
Mengzhuiwan	11.07	1.19	18.35	1.21	0.02	32.21	1.45	26.38	1.16	−0.29	0.73	−0.14
Duyuanjie	9.13	0.98	19.68	1.30	0.31	37.90	1.70	22.98	1.01	−0.69	−2.19	−0.19
Wangjiang	13.16	1.41	20.22	1.33	−0.08	43.49	1.95	37.29	1.64	−0.32	0.43	−0.20
Yulin	14.23	1.53	22.04	1.45	−0.08	38.00	1.71	31.28	1.37	−0.33	0.55	−0.21
Simaqiao	5.24	0.56	8.25	0.54	−0.02	27.14	1.22	18.14	0.80	−0.42	−2.99	−0.22
Fuqin	7.72	0.83	11.79	0.78	−0.05	36.55	1.64	28.48	1.25	−0.39	−2.00	−0.22
Xian	13.45	1.44	18.47	1.22	−0.23	34.93	1.57	27.69	1.21	−0.35	−1.11	−0.29
Hehuachi	5.50	0.59	6.71	0.44	−0.15	19.85	0.89	9.28	0.41	−0.48	−4.68	−0.32
Taisheng	9.68	1.04	13.31	0.88	−0.16	34.04	1.53	23.00	1.01	−0.52	−3.71	−0.34
Chunxi	10.04	1.08	12.52	0.82	−0.25	34.22	1.54	17.96	0.79	−0.75	−6.89	−0.50
Tiaosanta	24.24	2.60	29.47	1.94	−0.66	42.30	1.90	32.06	1.41	−0.49	−2.51	−0.58
Yanshikou	11.26	1.21	8.98	0.59	−0.62	36.86	1.66	23.03	1.01	−0.65	−8.06	−0.63
Inner City	9.86	1.06	18.51	1.22	0.16	32.23	1.45	28.78	1.26	−0.19	2.63	−0.01
Jinjiang District	8.52	0.92	18.12	1.19	0.28	22.09	0.99	27.26	1.20	0.20	7.38	0.24
Qinyang District	9.30	1.00	19.49	1.28	0.28	28.01	1.26	29.72	1.30	0.05	5.95	0.16
Jinniu District	7.73	0.83	13.26	0.87	0.04	20.36	0.91	21.21	0.93	0.02	3.19	0.03
Wuhou District	12.77	1.37	16.52	1.09	−0.28	23.67	1.06	21.55	0.95	−0.12	0.82	−0.20
Chenghua District	7.89	0.85	9.75	0.64	−0.21	19.03	0.85	18.45	0.81	−0.05	0.64	−0.13

(continued)

Table 3.1 (continued)

	Population over 6 and with university degree					Administrators/managers and professionals/technicians						Composite change 2000–2010	
	2000		2010		Change in LQ	2000		2010		Change in LQ		Absolute %	LQ
	%	LQ	%	LQ		%	LQ	%	LQ				
Main Urban Districts	9.31	1.00	15.19	1.00	0.00	22.26	1.00	22.79	1.00	0.00		3.21	0.00

Source Computed based on the population censuses of Chengdu

professionals and managers in these two years (LQ = 1.45 in 2000 and 1.26 in 2010). Not merely, across the entire inner city, societal restructuring was moderate throughout the 2000s. In the second-to-last column of Table 3.1, the composite index of the absolute percentage change was calculated by subtracting the arithmetic mean of the absolute percentage changes of the university-educated population and the professionals and managers in 2000 from the value for 2010. According to this composite index, the share of the high-status social class in the inner city expanded by a mere 2.63%, equivalent to, however, a decrease of 0.01 in the location quotient. When the campus relocation in the two university areas is not considered (Jianshelu and Wangjianglu), the inner city enhanced its competitive advantages in attracting a highly educated population over the decade. However, the comparative advantage of the inner cities in retaining professionals and managers decreased from 2000 to 2010. These indicators suggest a fundamental *status quo* of the inner city in China today. Multiple forces could drive inner-city social geography change. These forces could work together to reshuffle the inner-city social structure. Examples of such forces might include demographic densification, depopulation as a result of land-use transformation (e.g., from residential to commercial land use and public facilities), the growth of the working class due to migration and employment restructuring, and gentrification. In addition, urban expansion at the outlying areas of the city could prompt the inner city deconcentration of professionals and managers, even though during the 2000s the inner city continued to be attractive to highly paid employees.

Based on this background, two contradictory trends have dominated the pattern of societal restructuring at different locations over the decade. Eight subdistricts in the inner city of Chengdu are identified as having experienced intensified gentrification throughout the 2000s, while four became less gentrified. The share of middle-class residents declined in the other 19 subdistricts (Table 3.1). The population holding a university degree broadly increased across almost all subdistricts, benefiting from the expanded opportunities for university education for youth. However, measured by the LQ, the locational advantage for university graduates increased in 20 subdistricts but declined in the other 11. In addition to six subdistricts, a majority of the inner-city locations were affected by a relative decline in the LQ of high-ranking employees. Soaring employment in low-skilled services, enterprise reform and the de-concentration of residents could have limited the increase in the portion of professionals and managers. According to the analytical results, urban transformation in the inner city was asymmetrical and seldom evenly distributed across different areas. Gentrification stood at an extreme of the asymmetrical change in space and society that was inverse to the influx of the working class to ungentrified neighborhoods.

Finally, gentrification was inchoate in the inner city of Chengdu during the 2000s, but it soared in specific locations. Table 3.2 shows that over the decade, as the inner city gained 163,695 inhabitants with a university degree, it lost 10,530 professionals and managers. This was equivalent to a decrease of 0.01 in the LQ of the middle class in the inner city (Table 3.1). The eight core-gentrified subdistricts attracted 61,796 residents with a university degree and 11,620 high-ranking employees. Neighborhood redevelopment displaced 88,651 persons with less than a secondary education and 12,880 low-paid workers. The four less-gentrified neighborhoods observed

Table 3.2 Population change in the inner city of Chengdu, 2000–2010

	Change in the middle-class population[a]				Change in the working-class population[b]			
	By education		By occupation		By education		By occupation	
	No	%	No	%	No	%	No	%
Core gentrified	61,796	269.91	11,620	27.61	−88,651	−31.35	−12,880	−11.25
Less gentrified	22,021	131.23	1000	2.94	−44,596	−22.51	6890	9.18
Ungentrified	79,878	62.3	−23,150	−13.48	−54,612	−6.52	68,750	20.65
The inner city	163,695	97.5	−10,530	−4.24	−187,859	−14.25	62,760	12.01

[a]The middle-class population based on educational attainment includes the population with university and graduate degrees; the middle-class population based on occupational classification includes managers and professionals
[b]The working-class population based on educational attainment refers to the population with a secondary education and less; the working-class population based on occupational classification includes agricultural, production and commerce and trade service workers
Sources Computed based on the population censuses of Chengdu

a relatively small volume of change over these years, with a loss of 44,596 less-educated residents and an actual increase of 6890 low-paid occupations. By contrast, an extra 68,750 low-paid workers moved into the 19 ungentrified subdistricts, while the less-educated population generally declined in the inner city, including the 19 ungentrified subdistricts.

3.2   Spatial Manifestation of Gentrification

Based on the gentrification indices, this study mapped the geography of gentrification from 2000 to 2010. A brief introduction to the backgrounds of the five urban districts supports the understanding of this geography (Fig. 1.1). Located on the north side of the city, Jinniu District contains a significant amount of the land and housing occupied by the Bureau of Railways, which is affiliated with the central government, and by construction companies owned by the provincial government. Until 2000, Chenghua District in the east was one of Chengdu's major industrial districts; it contained large-scale manufacturing enterprises owned by the central and provincial governments. Adjoining the south of Chenghua District, Jinjiang District also held a number of manufacturing enterprises until a spatial plan initiated in 2007 attempted to convert the manufacturing base into a financial center. In addition, the traditional commercial and business center of the city is located in the district.

Wuhou District in the south contains several of the city's universities, and the earliest wave of real estate development began in this district in the early 1990s.

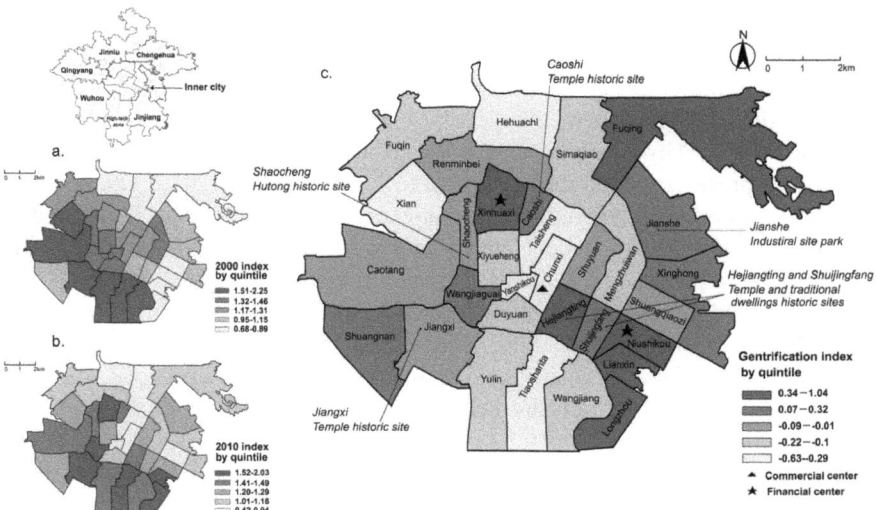

Fig. 3.1 Changing social status of the population in the Chengdu inner city, 2000–2010: **a** Social status in 2000. **b** Social status in 2010. **c** Changes in social status, 2000–2010. *Source* Drawn by the author based on the population censuses of Chengdu

After 1996, a high-tech zone was developed in four subdistricts in the southern part of Wuhou District. Currently, these subdistricts continue to be affiliated with the administrative division of Wuhou District but are directly managed by the municipal government.[4] Qinyang District occupies the western end of the city and has long been a purely residential area; it contains several public institutes and many of the cultural resources in Chengdu. As introduced in Introduction, the inner-city extent is defined as the areas within the Second Ring Road, while the downtown area is located at the geometric center of the city. Currently, the downtown area functions as the cultural and business center of the city, aggregating the city-level museum, library and theater.

The three maps in Fig. 3.1 depict the changing social status of the population in the inner city of Chengdu from 2000 to 2010. The 2000 and 2010 indices for the subdistrict population social status were calculated based on the arithmetic mean of the location quotients for the population with a university degree and professionals and managers. According to Fig. 3.1a, in 2000, the highly educated and highly paid populations tended to be located in Wuhou and Qinyang Districts.

---

[4] Since 2011, the high-tech zone in this map was administratively rescaled to depict the new urban district (Tianfu New District) at the south of the old city. Tianfu New District covers 1578 km², more than two times the area of the old urban districts. This study compared the socio-spatial change in the inner city with that of the five old urban districts, including the four subdistricts originally established within Wuhou District. However, the study excluded the new urban district, for the establishment of a new urban district can generate influential conditions that are distinct from those in old urban districts.

Some of the subdistricts in Jinniu District were also characterized by this middle-class concentration, which could be a result of the high-ranking workers occupying the large-scale work units in this district. In contrast, Jinjiang and Chenghua districts generally showed low settlement levels by high-status residents; this result could have been impacted by the dominance of the manufacturing base in the two districts. Figure 3.1c presents the gentrification quintiles from 2000 to 2010. The first two quintiles include the 12 neighborhoods that experienced either intensive or modest gentrification during this decade. Notably, gentrification elevated the socioeconomic upgrading of Jinjiang and Chenghua districts, as evidenced by the fact that eight of the gentrified neighborhoods were located in the two districts. In particular, at the southeast corner of the inner city, the five subdistricts alongside the river, which belong to Jinjiang District, had the most rapid upgrading in their social structure. The three gentrified subdistricts in Chenghua District were located along the outside of the First Ring Road.

Although not necessarily the most gentrified areas, all four designated historic sites and one industrial heritage site that has been successively renovated since 2000 were adjacent to the places favored by potential gentrifiers in the 2000s. Nevertheless, the localities of the two financial centers also represented a concentration of gentrifiers in this period. Until 2010, the highly educated and highly ranked populations were equally distributed across Qinyang, Wuhou and Jinjiang districts. After a wave of upgrading, however, Chenghua District still held a low proportion of the high-status social class residents, while the inner city of Jinniu District completely lost its advantage in retaining middle-class residents. In addition, the neighborhoods with the lowest ratio of middle-class residents, which are marked by the last quintile in Fig. 3.1a, b, tended to shift from the outside of the inner city in 2000 to the city core in 2010.

Gentrification geography thus resembles "islands of renewal" [2, p. 71], most clearly expressed in 2000 in a minority of working-class neighborhoods. However, such islands do not sit alongside inner-city seas of decay. China's inner cities have not yet witnessed urban blight. Specifically, the gentrification process has proceeded along multiple lines representing forces and processes that together promote radical urban transformation and reshuffle the inner-city social structure. Remarkably, tertiary industrial development and rapid urbanization have driven a working-class increase in inner-city low-skilled services. Thus, in the developing city of Chengdu, gentrification is not necessarily a dominant urban process. It forms part of the larger process of urban transformation. The transformation is asymmetrical and rarely evenly distributed. Gentrification is the extreme result of the asymmetrical change in space and society, denoting a unique and countertrending process. Unlike gentrification in postindustrial cities, which is characterized by the ascendancy of inner-city change in line with postindustrial transformation, the process in China may have relied on the leveraging of innovative urban practices to promote economic and social upgrading. The distinctive realities of gentrification in China require a new explanation of its meaning given the historical trajectory of urban and social change, representing a diversion from the postindustrial thesis.

3.3 Correlations Between Gentrification and Socio-spatial Transformation

To interpret the correlations between gentrification and various aspects of socio-spatial change, this study relies on nine sets of indicators to describe the social and physical attributes of an inner-city subdistrict: education, demography and household status, household registration status, industrial employment, occupation, housing tenure type, household rental expenditures, locational characteristics, and housing conditions. Three variables were assigned for each indicator, indicating one attribute of a place in 2000, one in 2010 and the value change over the decade. Three sets of coefficients were then evaluated by separately correlating the three sets of variables with the gentrification index (Table 3.3). The coefficients between the gentrification index and the independent variables in 2000 reveal those locational attributes that predict a high likelihood of middle-class occupancy. Associations between gentrification and the indicators describing the changing characteristics of a subdistrict in the 2000s explain the social and physical dynamics that delineate the gentrification process. The last set of coefficients indicates the association developed by 2010 between gentrification and the new socio-spatial attributes of a location.

Table 3.3 establishes simple correlations among the groups of variables with regard to gentrification.[5] Only the demographic and household characteristics presented no connections to gentrification. The household registration system continued to impede the right of disadvantaged groups to the city, reflected in the connection between gentrification in the 2000s and the percentage change of rural–urban migrants (-0.683). However, gentrification and the proportion of all migrant types exhibited no significant connection; this result indicates that, apart from rural–urban migrants, the arrival of interurban migrants and transnational populations, who are usually of higher socioeconomic status, could reduce the correlation. Thus, after industrialization, notably in the coastal cities of China, the increase in the proportion of highly educated migrants in the inner city is expected to have a positive effect on gentrification and most likely would alter the coefficient direction between gentrification and the migrant proportion in a location.

---

[5] Some of the occupational data on Jinniu District are missing, and they have been replaced by series means. Yanshikou is a subdistrict in which commercial land use exceeds 50% of land construction. Shuangnan Subdistrict was newly built on a greenfield site in 2000. The two cases are deemed to be not perfectly subject to the population of the gentrification samples. The outliers emerging in the two subdistricts were replaced by series means. Another set of unusual values emerged in the case of Jianshelu Subdistrict, which had the fastest growth in rental prices (14 times). Based on a field analysis, the study excluded the possibility of data error, but the large share of public rental and student housing in Jianshelu in 2000 could be the cause. To reduce the rate of information loss, the study replaced the outliers with the minimum/maximum values following the values for Jianshelu.

Table 3.3 Simple correlations for 2000, 2010 and 2000–2010 change against the gentrification index in Chengdu (N = 31)

	Correlations		
	2000	2010	Changes
Education			
Primary educated population	0.525	−0.576	−0.762
Secondary educated population	0.258	−0.535	−0.698
Population with a college degree	−0.459	0.524	0.737
Population with a university or graduate degree	−0.517	0.457	0.857
Demography and household			
Average age	0.156	0.078	−0.060
Female population	−0.042	−0.125	0.005
Average family member	−0.163	−0.001	0.075
Single-person household	0.095	0.126	0.227
One generation household	−0.164	−0.269	−0.156
Two generation household	0.186	0.314	0.124
Three generation household	−0.055	0.063	0.112
Household registration status			
Rural hukou population	0.086	−0.490	−0.683
Migrants	0.065	−0.342	−0.305
Industry			
Manufacturing workers	0.555	0.514	−0.530
Retail service workers	−0.088	−0.375	−0.256
Personal service workers	−0.468	−0.581	−0.341
Producer service workers	−0.377	0.261	0.635
FIRE workers	−0.275	0.468	0.644
Business service workers	−0.341	−0.207	0.202
Public service workers	−0.257	0.073	0.601
Public administration workers	−0.092	−0.211	0.148
Social service workers	−0.272	0.154	0.595
Occupation			
Managers	−0.262	0.439	0.578
Enterprise directors	−0.272	0.545	0.628
Managers in public institutions	0.344	−0.005	−0.311
Managers in government sectors	−0.340	−0.045	0.330
Professional and technical workers	−0.413	0.496	0.866
Economic business personnel	−0.246	0.305	0.350
Financial business personnel	−0.368	0.304	0.623

(continued)

Table 3.3 (continued)

| | Correlations | | |
	2000	2010	Changes
Engineers and technicians	−0.314	0.260	0.666
Teaching personnel	−0.290	0.382	0.502
Literature and art staff	−0.478	0.069	0.454
Clerk and related workers	−0.276	−0.260	0.060
Commerce, service trade personnel	0.070	−0.493	−0.692
Production workers	0.552	0.135	−0.529
Tenure type			
Self-built house owners	0.470	0.183	−0.405
Public tenants	0.519	−0.159	−0.449
Private tenants	−0.159	−0.546	−0.477
Owner occupants of commercial apartments	0.048	0.693	0.700
Subsidised owners	−0.598	−0.415	0.219
Owner occupants of price-controlled housing	0.452	−0.086	−0.345
Expenditure on housing			
Average monthly rent	−0.452	0.124	0.530
Locational character			
Standard of middle school		0.420	
Distance to master-planned financial and business centre		−0.441	
Distance to master-planned historic and cultural sites		−0.369	
Distance to traditional commercial and business centre		0.117	
Commercial land-use		−0.171	
Housing condition			
Households in housing built after 1990	0.283	0.515	0.500
Average floor space of building (per population)	−0.294	0.072	0.311
Households with independent bath	−0.462	−0.251	0.362
Households in building over 7 stories	−0.086	0.307	0.498
Households in reinforced concrete building	−0.240	0.426	0.068

Note Variables for social class in 2000 and 2010 represent the percentage status, except average age, monthly rent and floor space per capita, which are measured by absolute number. The value changes of the variables from 2000 to 2010 are calculated by the absolute percentage point change except for rental change and floor space change, which are defined as the growth rate. The level of middle schooling is defined as the educational level of middle schools that accept students from the sub-district. Middle schools are first scored based on three ranks: 2 = Nationally designated; 1 = Provincially designated; 0.5 = Municipally designated. The educational level is calculated by multiplying the rank score by the number of middle schools available to residents in the sub-districts. The locations of sub-districts relative to the most adjacent financial and business centres/historical and cultural sites are divided into four levels: 1 = within 0.5 km; 2 = 0.5–1 km; 3 = 1–1.5 km; 4 = 1.5–2.5 km. d. The locations of sub-districts relative to the city-level commercial and business centre are divided into four levels: 1 = within 0.5 km; 2 = 0.5–1.5 km; 3 = 1.5–2.5 km; 4 = 2.5–3.5 km
Sources Computed based on the population censuses of Chengdu, and spatial planning materials provided by the Urban Planning and Research Institute of Chengdu

3.3.1 State-Facilitated Industrial Transformation in the Inner City

The second column of Table 3.3 shows the correlation coefficients between gentrification and the independent variables in 2000. In 2000, new middle-class residents primarily arrived in manufacturing neighborhoods, as evidenced by the positive correlation between only the gentrification index and manufacturing employment (0.555) among all categories of the industrial population. The last column of Table 3.3 enumerates how the gentrification process is dependent on the aspects of socio-spatial change in a subdistrict from 2000 to 2010. As manufacturing jobs declined (−0.530), producer (0.635) and public services (0.601) increased, accounting for the social structure upgrading in a subdistrict. Specifically, considering the categories of the service sector, connections were principally established by the growth of employees in the FIRE (0.644) and social service (0.595) sectors. Similarly, among all professionals and technical personnel types, financial business workers (0.623) grew by one of the highest coefficients with the upward mobility of the social structure in place.

The correlations show that throughout the 2000s, gentrification in Chengdu was accompanied by deindustrialization and the development of financial industries. The locations occupied by manufacturing workers constituted the main gentrification battlefield in this decade. When the low-income production workers in manufacturing faced direct displacement (−0.529), the low-income commerce and service personnel also suffered from exclusionary displacement (−0.692).

The deindustrialization process, however, was fueled by government strategies promoting industrial transformation in the inner city. According to the *2009 Chengdu Yearbook*, from 2002 to 2007, the municipal government launched programs to relocate state-owned manufacturing enterprises in the Chenghua and Jinjiang districts to the eastern outskirts of the city while allowing for land reinvestment. The manufacturing enterprises in Chengdu were often located in *danwei* compounds, which provided the low-paid production workers with affordable rental housing owned by the companies. Nevertheless, not all of the former manufacturing bases turned out to be frontiers of new investment and social upgrading. For example, Shuangqiaozi and Mengzhuiwan in Chenghua District, containing a few declining manufacturing enterprises and their *danwei* compounds, experienced the clustering of low-income migrants and continual disinvestment during the 2000s. In contrast, the zones including the three gentrified subdistricts at the southeast corner of the inner city (i.e., Hejiangting, Shuijingfang and Niushikou) were developed as a new financial center in Chengdu, and the district government designated this area the Wall Street of Western China. As a result, these areas achieved thorough industrial advancement and landscape reconstruction in the late 2000s, resulting in the highest gentrification level.

Considering the industrial and occupational variables, the 2010 coefficients, listed in the third column of Table 3.3, were slightly different from those between the gentrification and independent variables for the changes from 2000 to 2010. Notably, a positive dependence continued between gentrification and the share of manufacturing workers (0.514) in subdistricts in 2010. Meanwhile, except for the FIRE sectors (0.468), no significant association was found between gentrification and a locational advantage for developing producer and social services. However, considering the occupational population, places that experienced social and physical upgrading were ultimately dominant in accommodating high-ranking employees in 2010.

The difference in the two correlation sets reinforces the idea of an incomplete transformation from manufacturing to knowledge-based service industries in gentrified neighborhoods. The gentrified neighborhoods were found to have a preponderance of high-status occupations but a certain degree of social mixing in industrial employment by 2010. One inference is that factory removal and resident relocation might have caused a high degree of retreat among the low-paid production workers but not necessarily among the high-ranking manufacturing population. Moreover, state-led factory removal was an inducement to gentrification but not a sufficient condition for it. The growth of gentrification relied on the successful generation of locational advantages in working and living environments that were attractive to high-income consumers from diverse sectors.

In addition to the development of knowledge-based industries, the transition from public to private sectors following China's economic marketization has been mirrored in its gentrification patterns. Among the three manager categories, an increase in enterprise directors was strongly correlated with inner-city settlement by the middle class (0.628), followed by an increase in managers of party/government departments (0.330). Managers of public institutions, however, presented an opposing influence on social upgrading in place (-0.311). Nevertheless, while the agglomeration of social services (e.g., education and healthcare) predicted the presence of gentrification in a subdistrict, the percentage change in public administration employees showed no relation.

This study thus reveals that state-led urban strategies for socioeconomic transformation in the inner city stimulated gentrification in Chengdu during the 2000s. This transformation was gradual rather than well established. As a result, instead of a full-fledged service class, professionals and managers, who constituted the main cohort of gentrifiers, came from different industrial sectors. These high-ranking employees preferred living places with favorable socioeconomic environments.

3.3.2 Housing Privatization

Table 3.3 shows a strong correlation between social upgrading and the increase in the owner occupancy of commercial apartments in a subdistrict (0.700). The 2000 and 2010 census data offer information about the household tenure composition at the subdistrict level. In both years, six housing tenure types are categorized according to the housing supply system in China: public tenancy, private tenancy, owner occupancy of self-built housing, owner occupancy of price-controlled housing, owner occupancy of originally publicly owned housing and owner occupancy of commercial apartments. The owner occupancy of previously publicly owned housing refers to residents who have purchased collective or public housing using government or *danwei* subsidies and is considered in this thesis to be subsidized housing. The owner occupancy of price-controlled housing was a product of the early stage of housing reform in the late 1990s and early 2000s. At that time, the government attempted to improve housing quality while maintaining its affordability. Self-built housing refers to housing built by households per se, often referring to rural housing and shanties built by either former villagers or existing urban residents.

In 2000, gentrification tended to appear in places with a large share of public tenants (0.519), self-built housing (0.470) and price-controlled housing (0.452). Those locations, dominated by subsidized owners, attracted less new investment and fewer middle-class newcomers (−0.598). Neighborhoods with a larger share of public tenants and self-built housing were usually located on land parcels managed by *danwei*s, municipal housing authorities or village collective organizations. The findings thus reinforce the influence of state-facilitated gentrification in China. In the 2000s, gentrification showed a tendency toward the polarization of homeowners in a subdistrict, as evidenced by the loss of households across all tenure types except the owners of commercial and subsidized apartments (0.700 and 0.219). Beyond the direct displacement of self-built homeowners (−0.405) and public tenants (−0.449), the decrease in private tenants is also related to the process of class replacement (−0.477), even though in 2000, the condition of private tenants in a neighborhood had no relevance to establishing the course of gentrification. Thus, in 2010, the gentrification process over the previous decade held a significantly positive correlation only with the proportion of households purchasing apartments in the free market. In particular, the process consolidated the socio-spatial segregation between owner-occupants of commodity housing (0.693) and private tenants (−0.546). The structural change in tenure is also reflected in an increase in actual expenditures on monthly rental fees by households in a neighborhood.

The strong coefficient between gentrification and the homeownership rate ultimately exposes the force of housing commodification and consumption underlying social geographical change in the inner city. In the context of rejuvenating private consumption in post-socialist China, the connection highlights the complexity of the social gentrification process in a transitional society. Beyond simply capturing a process of class replacement and displacement, the gentrification process in a Chinese city should unfold simultaneously with social change sourced from individual housing consumption and tenure change. Broadly embedded in the social transformation from a socialist to a market society, this social change is concerned with, on the one hand, the class-related behaviors and identities of high-income consumers encouraged by the development of commodity housing in the inner city. On the other hand, this social change is concerned with the circumstances and potential inequalities experienced by the current low-income households due to housing privatization. An explanation of the gentrification process needs to incorporate the relevance of spatial commodification and consumption in social class change.

3.3.3 The New Urbanism

Indicators of housing conditions and locational characteristics were assigned to define the place-making of gentrified neighborhoods. In the 2000s, the percentage change in housing built after the 1990s (0.500) and of residential buildings higher than seven stories (0.498) presented stronger connections with the gentrification trajectory than the share of housing with independent bathrooms or floor space per capita. This finding stresses that the specific housing form, more than the improvement of housing quality, served as a gentrification sign in the 2000s in Chengdu. Specifically, high-income consumers are inclined to reside in newly built, high-rise apartments in the inner city.

Locational characteristics are defined by the distance to financial, business and city-level commercial centers. Place advantages are measured based on the accessibility and quality of cultural, educational and commercial facilities. This study reveals that locations proximate to master-planned financial centers (-0.441) and historic and cultural sites (-0.369) were likely destinations of middle-class inhabitants during the 2000s (Table 3.3). This connection reinforces the argument that gentrification and financial industry development are tightly related. The five historical and cultural sites only rarely include residential buildings, but they include historic buildings and brownfields that were renovated and commercialized during the 2000s (Fig. 3.1c). Rather than acting as gentrifiable properties, cultural renovation for the purposes of tourist development and commercialization served as inspiration for further residential development and gentrification. In this case, these places are often less gentrified than the surrounding areas (e.g., Shaocheng vs. Wangjiaguai, Shuijingfang vs. Niushikou, and Jiangxi vs. Shuangnan).

Additionally, good schools at the primary and middle levels attract new capital and consumers (0.420). In Chengdu, admission to primary and middle schools in an area is contingent on residential proximity and the household registration location. Although the municipal government has attempted to balance geographical differentiation in educational quality by opening admission to high-standard schools to a certain number of nonlocal residents, residents living in the immediate vicinity of high-standard schools have substantially more educational choices and lower enrollment costs. Finally, gentrification has no obvious relevance to the commercial facilities in these neighborhoods. In fact, the places with the highest proportions of commercial land use presented the lowest probabilities of gentrification and even displayed poverty concentrations, which might be partly the result of the large share of low-paid employment in the consumer service industries in districts with extensive commercial land use. For instance, in Yanshikou, more than half of which is apportioned for commercial land use, the socioeconomic profile experienced downward mobility from 2000 to 2010 (Table 3.1).

The gentrified landscape reflects a consumer preference for areas with cultural resources and a favorable social fabric. Moreover, it highlights the influence of the state-led creation of new urbanism in the inner city on the cultivation of consumer culture among high-income consumers. Since 2000, Chengdu has ushered in a period of rapid development, responding to national strategies and policies encouraging development in Western China. To date, four waves of large-scale urban renewal have been launched: in 1993, 2002, 2009 and 2012. While urban reconstruction before 2000 began the waterfront area renovation, the two waves of urban renewal in the 2000s, led by Secretary Li Chuncheng, notably guided the renovation of inner-city historic sites (Shaocheng, Caoshijie, Hejiangting and Jiangxijie). Additionally, together with the relocation plan for factories that was formulated in 2007, the government has increasingly shifted its focus to upgrading not only the built environment but also, substantially, industries. These state-led projects overlap with the geography of residential gentrification, which explains the efficacy of state-led place changes in stimulating the spatial consumption and residential mobility of gentrifiers. The most recent wave of redevelopment was set in motion in 2012. Chenghua District, which still contains a large proportion of factories and *danwei* compounds of state-owned enterprises, has now become the new frontier of investment in high-end residential and commercial estate.

Similar to the new middle class's embracing of inner-city urbanism, China's new rich are stimulated by it in inner-city areas but in a unique form. This new type of urbanism has been characterized by a new housing pattern, good quality education, and enriched resources representing historical culture. The distinctive cityscape conveys the different causes and characteristics of sociocultural change in contemporary Chinese cities. Most notably, rather than a product of the cultural trend of a lifestyle change among the new middle class in quaternary industries, it is produced by the new market regime with support from the reformist state. It represents an imagined new modern landscape after the dystopia of Mao's envisaged socialist city. Correspondingly, it characterizes a cultural transformation toward a new Chinese modernity, wherein China's new consumers are at the cutting edge of cultural delivery.

The dependencies of gentrification on spatial commodification and newly built urbanism ultimately introduce the forces of production and spatial consumption into gentrification in the Chinese context. Based on public land ownership and the welfare housing system, the politico-economic elite has guided institutional change toward spatial commodification and the production of new landscapes in contemporary Chinese cities. socio-spatial production will cultivate new urban lifestyle types and consumer cultures. Meanwhile, gentrifiers have expressed a propensity toward new urbanism and homeownership in the inner city in their quest for landscape and neighborhood credentials—socially, culturally and economically. An explanation for gentrification must discern the social class dynamics induced by the interaction between the forces of production and consumption in the inner city.

3.4 Conclusion

Gentrification in post-socialist cities, as well as in other Southern cities that seem on a fast track to economic takeoff, appears as part of a time of profound urban transformation. Multiple urban processes are entangled with each other and lead to revolutionary changes in the social space. The urban reality directly renders gentrification in these southern cities less observable and comprehensible. Part One attempts to disentangle the thread of gentrification from the complex process of urban transformation by materializing the process and articulating it with major urban trends in the post-socialist era. We thereby generated three relevant aspects to understanding gentrification in China. These act as starting points to explain the practices involved in gentrification.

First, gentrification relies on urban development strategies enacted by local state actors. Because quaternary industries in post-socialist cities have developed only modestly, gentrification in the 2000s was concentrated in fewer localities in the Chengdu inner city, in particular, in manufacturing neighborhoods that were planned and redeveloped to kick-start a new growth cycle. Moreover, the tendency of urban and social change could be juxtaposed against wider urban restructuring in the inner city, which adapts to the socioeconomic context of expanded low-skilled service workers. Gentrification reflects the most innovative city practices and tends to transcend the developmental trajectory and catalyze rapid social and economic advancement.

Second, the new urbanism created by state-led strategies became the main force attracting gentrifiers into the renovated inner city. State-led urban strategies could lead to new residential, retail or office spaces, with the larger objectives of economic restructuring, real estate development or urban beautification. Such strategies, in general, engender new symbolic meaning in urban places and lifestyles, which substantially attracts inner-city gentrifiers. Moreover, such new urbanism exemplifies the urban transformation in a commodity economy, along with China's transformation from a distributive to a market economy. The analysis demonstrates a strong dependency between social upgrading in a place and private housing consumption and the occupation of the newly built urbanism.

Last, together with the state and developers, a group of middle-class consumers constitutes an important force behind gentrification. In the socioeconomic context of an incomplete postindustrial transformation, the study rejects the assumption that a highly educated service class drives gentrification in China. Instead, it emphasizes the variety of gentrifiers when considering industrial sectors and production systems. In the context of the consumer revolution, the variety of gentrifiers is combined with new, place-based consumer cultures and habitus. The argument thus encourages an examination of the gentrification process in China, which is accompanied by intricate social change, amplifying the process beyond a straightforward struggle between two socioeconomic classes.

The above conditions permit an understanding of gentrification in China that moves away from being a direct result of the class invasion of inner-city locations or the expropriation of land value by developers and toward being a more complex nexus of the state-facilitated creation of new urbanism, the consumption of spaces by the middle classes, and the social and urban changes that occur as a result of the combination of these two activities. An explanation of state-facilitated gentrification should emphasize the interaction of these production and consumption forces, with a stress, as [9] insist, upon the role of the state.

References

1. Badcock B (1989) An Australian view of the rent gap hypothesis. Ann Assoc Am Geogr 79(1):125–145
2. Berry BJL (1985) Islands of renewal in seas of decay. In: Peterson PE (ed) The new urban reality. Brookings, Washington, DC, pp 69–96
3. Cai F, Park A, Zhao Y (2008) The Chinese labor market in the reform era. In: Brandt L, Rawski T (eds) China's great economic transformation. Cambridge University Press, Cambridge, pp 167–214
4. Clark E (1988) The rent gap and transformation of the built environment: case studies in Malmo 1860–1985. Geogr Ann Ser B Hum Geogr 70(2):241–254
5. Davidson M, Lees L (2005) New-build "gentrification" and London's riverside renaissance. Environ Plan A 37(7):1165–1190
6. Goodman DSG (2014) Middle class China: dreams and aspirations. J China Polit Sci 19:49–67
7. He S (2010) New-build gentrification in central Shanghai: demographic changes and socioe-conomic implications. Popul Space Place 16(5):345–361
8. Lees L (2003) Super-gentrification: the case of Brooklyn Heights New York City. Urban Stud 40(12):2487–2509
9. Lees L, Shin HB, López-Morales E (2016) Planetary gentrification. Polity Press, Cambridge, Malden, MA
10. Ley D (1996) The new middle class and the remaking of the central city. Oxford University Press, Oxford
11. Li C (2006) Characterising China's middle class: heterogeneous composition and multiple identities. In: Li C (ed) China's emerging middle class. Brookings Institution Press, Washington, DC, pp 135–156
12. López-Morales E (2011) Gentrification by ground rent dispossession: the shadows cast by large-scale urban renewal in Santiago de Chile. Int J Urban Reg Res 35(2):330–357
13. Lu X, Song G, Hu J (2007) Executive summary. In: Lu X (ed) Social structure of contemporary China. World Scientific Publishing, Singapore, pp 1–64
14. McKinsey (2013) Mapping China's middle class. McKinsey Quart 3. Retrieved June 29, 2016, from: http://www.mckinsey.com/industries/retail/our-insights/mapping-chinas-middle-class
15. Smith N (1996) The new urban frontier: gentrification and the revanchist city. Rutledge, New York
16. Walks RA, Maaranen R (2008) Gentrification, social mix, and social polarisation: testing the linkages in large Canadian cities. Urban Geogr 29(9):293–326
17. Wyly E, Hammel D (1999) Islands of decay in seas of renewal: housing policy and the resurgence of gentrification. Hous Policy Debate 10(4):711–771
18. Wyly E, Hammel D (2004) Gentrification, segregation, and discrimination in the American urban system. Environ Plan A 36(7):1215–1241
19. Zhou X (2004) Chinese middle class: reality or illusion? In: Jaffrelot C, van der Veer P (eds) Patterns of middle class consumption. Sage, London, pp 110–126

Part II
Production and Consumption of Gentrification

Chapter 4
Initiation

Based on the three conditions identified in the last chapter, this chapter explains the causes of socio-spatial upgrading in the inner city. It emphasizes the way in which the local state acts in spatial production through its orientation to cultural and economic institutional change related to urban construction. Understanding state power over land use is the starting point for scrutinizing the role of state played in gentrification in China. In the early days of the People's Republic of China, land in Chinese cities included state-owned land and private land held by private capitalists and individuals. Until 1982, public ownership of urban land had been formally established in the Constitution. At this stage, the government's power over land-use was mainly manifested in administrative rights, instead of the power over land-use planning. The city lacked planning, and the land was used inefficiently. The state allocated land to *danwei*s through administrative allocation or transfer, and the *danwei*s used the land free of charge and indefinitely.

The Constitution and the Land Law passed in 1988 formally established that land-use rights could be transferred. Since then land-use rights and ownership were separated. Land property rights included land ownership as well as land-use, land development, land leasing, land mortgage and other rights [79]. Ownership confers the most comprehensive power. Land-use rights are a kind of tradable property right that can be realized in two ways: leasing by the government or the former land user and allocation by the government.

Based on the system of the paid use of land, central and local governments exercise the two types of public power: land expropriation and land-use planning, which ensures the emergence of gentrification. Land-use planning and land expropriation are administrative powers that generally exist in many countries, which stem from the fact that land use has externalities [104]. To protect social and environmental justice and other public interests, the government needs to intervene in land use. But one feature of the land-expropriation right in China is the ambiguity of its exercise premise [99]. China's Constitution, the Property Law and the Land Law all make it clear that the rationality of land expropriation lies in public interests [53, Artical 10, 54, Artical 2]. However, these laws do not explicitly or convincingly define

© Springer Nature Singapore Pte Ltd. 2022   75
Q. Yang, *Gentrification in Chinese Cities*, Urban Sustainability,
https://doi.org/10.1007/978-981-19-2286-2_4

public interest. In view of the housing expropriation on state-owned land in cities and towns, the State Council promulgated the *Ordinance on Housing Expropriation and Compensation* on *State-owned Land* in 2011. The Ordinance specifies six circumstances under which housing and land can be **expropriated**. The fifth is the renovation of dilapidated housing. However, the Ordinance neither orients the land use after expropriation nor explains how to determine which land use is in line with the public interest. Nonetheless, the regulation clearly notes that land expropriation should be compensated because the right of expropriation directly deprives the user's land-use right. Therefore, state power over land expropriation was characterized by compulsion and compensation, and the government maintains accountability for the preconditions of expropriation [71].

The major characteristics of state power over land-use planning in China lie in the extension of rights. Spatial planning in China includes overall land-use planning under the supervision of the land department, the national plan of the main functional zones under the supervision of the Development and Reform Commission and urban and rural planning under the supervision of the housing and construction department. The urban and rural planning system further includes the planning types—such as strategic plans, master plans, zoning and urban design—at different administrative levels. Since 2013, China has successively carried out the reform of "integrating multiplanning" and "territorial spatial planning" [44]. In 2018, the Ministry of Natural Resourses was established, which tends to unify the exercise of responsibility for the control of all national space uses. It further strengthened state power over land use. The planning system has created the broad connotations of China's spatial planning and its close relationship with social and economic development [13].

Gentrification is mostly related to the exercise of the planning power by local governments over urban land. In countries with a private land system, urban planning is essentially a way of regulating conflicts among private property rights or the contradiction between the exercise of private property rights and public interests [76]. *The Constitution* and *the Land Administration Law* in China do not prescribe any prerequisites for the exercise of planning rights in China. Urban planning is thus often used by local governments to achieve their own interests or purposes. The situation became pronounced after the reform of the tax distribution system in 1994, when local governments experienced mounting financial pressure and intensified competition [76]. The objectives of urban planning include not only the public interest but also prominent economic and political purposes or a balance between economic and social development [90, 92].

Meanwhile, the types of planning that assist local governments' control over land development are also varied. At the city level, statutory urban planning includes master and regulatory planning, while nonstatutory planning includes development strategic plan, city quality improvement plan, urban renewal, rural revitalization, cityscape plan and so on. These nonstatutory plans usually reflect the municipal government's urban development strategy at that time. Among these plans, the initiation an implementation of gentrification could be most impacted by state-facilitated planning related to urban image-building and the adjustment of regulatory planning to generate new land-use conditions.

In sum, the extended power over land-use has pushed the government to identify broader goals for spatial planning. The fact that public interest is ill-defined in the law leaves room for government discretion in land and housing expropriation. The result is that, on the one hand, the government may conduct property expropriation under the guise of planning types, which, in general, is for the public interest. One example of the combination of these two rights is the large-scale expropriation of old houses and collective land caused by the government's urban renewal and new town plans. The state power over urban planning, on the other hand, also limits the mode of land use by market actors and determine the social obligations of the subsequent holders of land-development rights. The two types of power thus underpin state intervention on the making of a new cultural urbanism, the implementation of spatial commodification and concomitantly re-housing the affected residents. This chapter will discuss the first two sets of state practices.

4.1 New Cultural Urbanism

After decades of urban (re)development, large Chinese cities now present various modern, late-modern and postmodern landscapes with corresponding forms of sociability. When passing through a historic commercial district, one may see department stores with clumsily added exotic markings and young people and migrants crowded into central squares who are shopping, walking, searching for local cuisine or seeking retail job opportunities. Close to historic commercial centers, towering buildings with light gray glass, clean walls and clean lines rise abruptly above uniformly multistory, drab concrete buildings that represent the legacy of Mao's China. Within high-rise towers, well-known global luxury brands are for sale on the first floor, with IMAX theatres above and upscale restaurants on the roof decks. Stylish young people and elegant women with suited men pass in and out of the buildings, while others are only afforded glimpses inside. People know well where they are allowed to go. Near the towering city centers, stark, low-density districts present Chinese cultural histories to tourists. Buildings are two or three stories tall with postmodern combinations of sloping roofs and with glass and wood-grained walls of a deep-gray hue with Chinese characters. This style demonstrates how cities are positioned between rapidly modernized cityscapes and their celebrated histories. Nearby, one can find neighborhoods with bars, clubs, and coffee shops (with names reminiscent of the Soho districts in Hong Kong, New York and London); the tallest and most expensive condominiums; five-star hotels; and westerners jogging outside. When leaving the city center and approaching a new urban district, one rarely finds typical suburban landscapes. Instead, rows of buildings tower along broad avenues. Certain areas have shopping malls, large complexes and stadiums, and landscapes reminiscent of La Défense. As one travels away from densely urbanized districts, luxury suburban communities finally come to the fore, not next to large shopping centers but with everything one needs within gated communities, including malls, flagship stores selling global brands, laundry services, restaurants, theatres and even churches.

These landscapes are indicative of modern urban China, i.e., renovated and newly built large cities in the post-socialist era. In the 1960s and 1970s, classical gentrification in advanced capitalist cities was developed in the advent of postindustrial and postmodernist urbanism, initially driven by cultural pioneers and then championed by the new emerging middle class. In objecting to the supremacy of economic values and formalistic cultures under Fordism and mass production, the new middle class's engagement with historic preservation, neighborhood construction, and political reform reflected a trend of social liberalization and cultural innovation [31].

Different from gentrification that has its genesis in a postmodernist cultural shift, this section shows how state-led gentrification originating from a nationwide cultural project is accompanied by an ascendancy of modernist consumer cultures and urban lifestyle that condemns alternative lifestyles to obsolescence. Drawing on the concept of city branding, it decodes the cultural message in the trajectory of cityscape-making and representation resulting from state-facilitated urban redevelopment in Chengdu.

4.1.1 Anti-urbanism and Deaestheticization

Mao Zedong had a fundamental influence on Chinese modernity and social formation. Throughout the 1950s and the later period of the Cultural Revolution, Mao represented a trend of anti-urbanism, encouraging an austere lifestyle with mass production and subjugating individual autonomy to national development [6, 47, 48]. The trend of anti-urbanism left Chinese cities underinvested in their built environment and deaestheticized for the impoverished masses. In 1985, more than two-thirds of the housing floor area in 323 Chinese cities was under the jurisdiction of work units, while municipal housing authorities were in charge of another 9% of the housing floor area [86]. To date, the socialist landscape continues to predominate in Chinese inner cities ([47]; Fig. 4.1). This section reviews the cultural legacy of the socialist cities managed under Mao's rule. It is beyond the scope of this chapter to trace their genealogy. The following brief review offers a glimpse into the social and cultural foundations from which gentrification has unfolded in the post-reform era by outlining relationships between place creation, lifestyles and class distinction.

The *danwei* system of post-1949 cities entailed the establishment of public housing allocation and management systems involving two levels of institutions and enterprises and the municipal government. The "production first, living second" doctrine generated an anti-design movement for housing. From Soviet expertise and notions of modernist typology, *danwei* housing is typically characterized by six-story, concrete blocks of a minimalist and standardized design. Additionally, widely constructed workers' neighborhoods, which accommodate the production workers of state-owned enterprises, resemble bachelor apartments but are of far inferior quality. The two- to three-story buildings within the workers' neighborhoods are referred to as barrel buildings (*tongzilou,* 筒子楼; Fig. 4.2). One nuclear family may crowd together

Fig. 4.1 The socialist and newly built landscape in Chengdu. *Source* Photo taken by author, 2016

Fig. 4.2 Workers' neighborhoods in Chengdu. *Source* Photos taken by author, 2016

in a single room of less than 20 m^2 without a bathroom. Rooms are positioned along corridors, and residents must cook within these corridors and share public toilets [35]. Within the *danwei* compound, private spaces are heavily compacted and intermingle with public spaces such as canteens, playgrounds and meeting rooms [45].

> According to the decision of the Construction Committee, "the new types of housing must give the workers maximum free time and energy for their social and cultural activities, must place at their disposal suitable leisure opportunities and facilitate the passage from an individualistic concept of housing to more collective forms. [2, p. 85]

Indeed, the government made preliminary attempts at residential spatial planning by importing Western concepts of the neighborhood unit. Meanwhile, architectural forms in Beijing at one time recovered traditional cultural elements through modernist buildings to create archetypes that symbolized Chinese socialism and nationalism [45]. Such aesthetic trends involved waves of eclectic experiments in building constructions, showing signs of Chinese and Western classicism [34]. However, these concepts were soon criticized and replaced by politicians' prioritizing of economically efficient construction methods. New housing and neighborhood construction types later absorbed the ideas behind Soviet housing industrialization and Mao's People's Communal System.[1]

Meanwhile, since 1957, governments have confiscated and socialized private property constructed before 1949. Such sites mainly included courtyard houses owned by capitalists in prerevolutionary China. Various types of courtyard houses are adapted to the geographical features of Northern and Southern China. While courtyard houses in Northern China are typically bungalows with spacious skylights, two-story houses with small skylights predominate in the south. Courtyard houses are often built back from crowded streets and positioned along small lanes (e.g., *hutong* in Beijing and *lilong* in Shanghai; [29]). After these houses were confiscated, municipal governments divided and rented these buildings to residents as public housing. The saturation of low-income workers and a lack of investment from municipal governments accelerated the degradation of those historic buildings. In 1968, a reformist committee of the Cultural Revolution was organized in Chengdu. After December of that year, the municipal government demolished the Qin Dynasty city walls and an iconic premodern Chengdu building located at the geometric center of the city, which had traditionally been used as an examination hall for scholar–officials and an administrative center.The area was then reconstructed into a plaza surrounding a sculpture of Chairman Mao. This spatial destruction targeting imperial China foreshadowed decades of urban modernization in Chengdu.

The landscape of *danwei* compounds embodies the sociocultural politics of Mao's China. The deaestheticization of spaces implies a cultural Maoist strategy that removes individualism and magnifies collective subjectivity to submit personal pleasure to national production. It shrinks the private domain of daily life and establishes a public domain in *danwei* compounds for socialization that is closely associated with the values of a national totality [2]. Nevertheless, collective consumption in *danwei* compounds results in a homogenous lifestyle and culture among cadres and workers. However, it would be a mistake to interpret *danwei* society as simply "classless" (see [85]). Professionals, managers, and technicians were lumped together with workers not due to their balanced social and economic statuses but due to the need for alliances in the production movement. Skilled and educated cadres included think tanks, assistants and performers of the socialist state and were role models for mass society. Workers were trained as completely socialist soldiers of production but still lacked political power in decision-making. This social condition is exemplified by the status of the working-class trade union in socialist China. The responsibilities of the trade union brought it into conflicts with the *danwei* administration, which ultimately raised a debate within the Party regarding the relationship between the trade union and *danwei* administrative authorities. On the one hand, the trade union

---

[1] Within the Jiangxi Soviet period, in 1932 Yanan, Mao established a centralized system of fiscal control and redistribution. However, in Yanan after 1939, the CCP government faced economic difficulties. Thus, the mass mobilization of production was initiated as a long-standing policy for boosting the economy. Separate production units were established to kick-start self-production and to allocate goods and welfare to members. This type of productive community was a predecessor of the People's Communal System, the lowest sociopolitical unit adapted to Mao's ideas of social organization and economic production [2, p. 47].

should be independent of the administrative system to focus on workers' affairs. On the other hand, central control is an essential principle of socialism, so separation implies a substantial error symbolic of capitalism. The latter argument won, while the former was denounced as departing from socialist ideology and doctrine [2].

From the 1980s to the mid-1990s, the state-led reconstruction of dilapidated buildings was set in motion in Chinese cities in the context of housing shortages and unsound built environments [17, 22]. This new wave of community construction inherited the compound style but improved community services and housing quality levels. Nevertheless, the central government continued to encourage the compact, affordable and economical construction of new residential compounds to achieve a level "between minimal and luxury" that was integral to the social objective of development toward a moderately prosperous society (*Xiaokang shehui,* 小康社会) in China [28]. In Chengdu, the municipal government started the first large-scale redevelopment program in 1993 to encourage waterway environmental improvements and infrastructure construction. The waterway improvements were completed in 1997 and were granted the UN–Habitat Scroll of Honor Award in 1998 [75]. The volume of building demolition from 1992 to 2002 accounted for 20 million square meters and involved moving 110,000 households to 24 newly built communities [33]. Most of the newly built houses thus remained basic in design and were contained within 60 m^2 for a nuclear family of three or four people, but the houses also separated functions and improved overall areas. Within these communities, uniform multistory apartments were arranged in rows with a public space positioned at the center ([102]; Fig. 4.3).

Housing and neighborhood construction led by the state in the pre-reform era had a fundamental influence on cultural change in the post-reform period. Urban reconstruction before the mid-1990s established the typology of gated communities; by fully repudiating traditional cultures and erasing the signs of the past, it expedited the secularization of Chinese culture and left a mass society with a minimal aesthetic consciousness and class distinctions in habitus. Ideologically and physically, the spatial practices and representations of this period flattened state-led socialization

Fig. 4.3 New residential community in Chengdu in the early 1990s. *Source* Annual Report of Chengdu, 1999

and cultural construction via urban (re)development in the subsequent years. After learning lessons from urban renewal projects in Shanghai and Beijing, an official from the Chengdu Housing Department recognized the following:

> After years of "massive destruction and reconstruction" (*dachai dajian*), we now consider protecting some old buildings, as you mentioned, such as Sichuan dwellings and Soviet buildings. The central Chengdu landscape has not changed in five thousand years. However, Chengdu has nearly lost its identity as a historic city. Very early on, the municipal government started to bulldoze quadrangle dwellings in the city. Soviet buildings above a certain size are mainly located in Chenghua District. However, most of them were demolished when the district government launched its livelihood projects for urban redevelopment and when the Chengdu municipal government decided to relocate secondary industries from the inner city to Chenghua District. (O22, official from the Urban–Rural Housing Department of Jinniu District)

4.1.2 Defining Global City Image

Rapid urban change in many large Chinese cities began in the mid–to-late 1990s. China's accession to the World Trade Organization and the national reforms of land-use system in 1988, tax distribution system in 1994 and housing system in 1998 together increased local governments' enthusiasm to promote urban development. In particular, regional and national central cities have successively put forward the objective of modernist and global city-making. Beijing's global city construction set in motion in 1991, together with the initiation of building new CBD, high-tech zones and old city renewal. To prepare for the Beijing Olympic Games, the municipal government formulated Beijing 2002 Action Plan. The plan was devoted to urban infrastructure construction, rural land acquisition, urban cultural facilities construction and other projects, so to form a showcase of a renowned, historical, cultural city and a large, modernized, international metropolis. At the same time, the master plan in 2004 set a two stage goals for the urban construction: to achieve modernization and become an international city by 2020 and to become a world city by 2050.

In 1990, the State Council announced the development of Pudong New Area in Shanghai [87, 88]. Shanghai Master Plan (1999–2020) put forward the goal of building Shanghai into an international metropolis, being a hub *of the world* economy, finance, trade and shipping. Throughout the 1990s, Shanghai hosted a variety of global events such as the Fortune Forum and APEC meetings, which greatly enhanced the recognition and understanding of Shanghai in the international communities. The government and the media endeavored to articulate the image of Shanghai with that of Western cities by, for example, the reuse of Shanghai's historical brand "Oriental Paris" or, the comparison of the cityscapes of Shanghai and Manhattan at the Fortune Forum [88].

Following Shanghai, Guangzhou proposed catching up with the four small dragons in Asia by 2015; Dalian strived to build "Northern Hong Kong" by 2020; Wuhan aimed to be "Oriental Chicago". Today, 182 cities have proposed building an international metropolis, accounting for approximately 27% of the 667 cities in the country. Large-scale urban renewal programs often followed the proposals of global city making. From the 1990s to the early half of 2000s, the mode of state-led urban renewal was dominated by demolition and construction, with the goals of the construction of infrastructure and public facilities, urban modernization, beautification and housing development.

A critical moment of urban change in Chengdu occurred in 2002 with the innovations of the former party secretary from 2003 to 2009. Secretary Li Chuncheng embraced theories of city marketing and paid specific attention to city branding and urban imagery. In 2001, when he was promoted to Mayor of Chengdu, Li launched a shantytown reconstruction project, which was located in the most-gentrified neighborhood: the Lianxin subdistrict (see Fig. 3.1c). The project was Mayor Li's first engagement with the city. Rather than merely encouraging demolition and reconstruction for environmental improvement, Li attempted to rebuild the urban image. In 2003, Li invited a famed Chinese director to shoot a city video entitled "Chengdu: Once you come you will not want to leave," representing the first account of Chengdu to the world. Told from an outsider perspective, the video stresses the sociocultural aspects of the city more than the city's economic culture. It presents a city with friendly residents, a comfortable and leisurely lifestyle and an active teahouse public life. Iconic spaces include the redeveloped Tianfu River and remaining imperial and religious spaces throughout the city.

Over the next several years, Li organized research and media reports on the branding and marketing of Chengdu. Based on the city's comfortable climate and long-established folk culture tradition (e.g., teahouse culture), research and media reports have constantly strengthened Chengdu's image as livable and inclusive while stressing its need for global acclaim. One of the influential promotional tactics for Chengdu is its "Fourth city" branding, denoting the city's status as having the potential for development akin to that of Beijing, Shanghai and Guangzhou. As early as 2000, the New Weekly, a Guangdong-based newspaper that is well known for its creative views on social change, published a full issue on why Chengdu is considered the Fourth City of China. Four years later, the Western China City Daily of Chengdu and the New Weekly held a forum in Chengdu that involved reviewing the Fourth City's development in recent years. During the forum, the chief editor of the New Weekly described his views on Chengdu as a leader of urban lifestyle change and cultural innovation in China.

> Economic capacity cannot represent the identity of the city that will not be measured using statistics. A city's charm concerns its lifestyle, taste, and aesthetics. Chengdu, a city in the Western region of China (known for a relatively low pace of economic growth), remains at the cutting edge of lifestyle creation. [80]

The Chinese Academy of Social Science, the top think tank in social science based in Beijing, also conducted a series of surveys to measure the efficacy of Chengdu's branding both internationally and domestically. A report showed that Chengdu receives little recognition globally but greater recognition domestically. It defines Chengdu's city identity as "an inclusive, open inland city with rich and distinctive historical cultures" [55, p. 105]. Chengdu is described as having the potential to become an "internationally notable cultural city," with the city's notable character-istics including its livable environment, enriched historical resources; amicable and spirited residents; comfortable lifestyle; and trendy, youthful aesthetics (pp. 106–108). However, the report also stressed that the level of cultural amenities remains relatively low and must be upgraded to international standards.

The city marketing strategy implemented various plans for creating model cities of environmental and social development (i.e., "National Hygienic Cities" and "National Civilized Cities") from 2002 to 2008, such as community development, street beautification and the removal of street vendors [74]. Cartier [6] focused on the National Civilized Cities program as an instructional program in which the state acts to provide knowledge on desirable commodities to buy, establishing a model for consumer citizenry. The civilization program, at least initially, was regarded as polit-ically necessary to unify society and balance the negative influence of the commodity economy and material consumption [5–7]. After failing in the first competition round for the National Civilized Cities contest in 2005, the municipal government of Chengdu and local media organized a public discussion and self-reflection on "the Gap of Chengdu to Civilization" [20]. This discussion ultimately became a social learning tool citywide.

> The city project may come to naught simply because of uncivilized conduct by individuals. We need everyone in the city, including citizens and migrants who work or do business in the city, to present themselves at their best and to obey social and moral norms. [77]

Meanwhile, after the administrative reforms of the 1990s, communities replaced work units as the smallest administrative areas. However, the city renewed the impor-tance of community-building for social control, following President Jiang Zemin (1989–2000), who described "community-building" as "approaches that promote social development, expand grassroots democracy, raise living standards and main-tain social stability" [3, p. 536]. The functions of residents' committees were extended from the realm of service delivery to include cultural publicity, social education, policing, etc. (e.g., "civilized citizens–quality education") [3]. Similarly, the central state spread moral discourse on "quality" (*shuzhi,* 素质 to motivate the self-improvement of dominant labor segments to facilitate the country's economic transition (e.g., young adults; [58]). Since then, older neighborhoods have been stig-matized as blighted spaces that are less civilized and subject to "control, redefine and transform" [5, p. 1526]. Local decision-makers view neighborhood renewal and spatial regulation as necessary to facilitate social civility, governance and the improvement of cultural taste among citizens (O33). Rural–urban migrants have clearly suffered from redevelopment due to the rampant removal of street vendors and the formalization of space.

> For urban governors, the old and declining places in the city are obstacles to urban develop-ment. The places are insecure, rife with social conflict and untouched by social governance. Floating labor, low-income citizens, and disadvantaged senior populations have occupied these kinds of places. They cause the duality of social spaces in the city and development difficulties. [11]

Based on these initiatives, Secretary Li promoted the most extensive urban redevelopment plan in Chengdu from 2002 to 2004. Over these three years, more than 5.8 million square meters of housing were cleared away, and 110,000 households were relocated [105]. During this period, a large-scale urban project called the "Structural Adjustment Planning of East Suburban Industrial Zone in Chengdu" resulted in a far-reaching impact on subsequent gentrification in Chengdu. The manufacturing area was built up during the early stage of China's industrial revolution as part of national industrial strategies. The industrial zone was primarily located on the east side of the inner city of Chengdu and developed along both sides of East Avenue. In the early 2000s, the region gathered more than 400 enterprises of different sizes— mainly in the electronics, machinery and metallurgical industries—with more than 150,000 employees [98]. As the city developed, the area formerly known as the eastern suburbs became the heartland of the city. In 2000, the national government established the great western development strategy. In 1995, the Chengdu master plan proposed the objective of urban modernization. The project was thus put forward of manufacturing relocation, which aimed to improve the urban environment and city image and adjust the urban functional structure. In 2002, the "Structural Adjustment Planning of East Suburban Industrial Zone in Chengdu" was finished.

Within this project, municipal and district governments have actively intervened in land use and prices after redevelopment. First, the plan proposed that vacated land would be mainly used for residential, public service and green space development. Then, the environmental renovation of riverside areas was a main project for advancing district quality. Third, the government proposed guiding developers to vacated areas by formulating attendant land supply policies. Finally, the government intervened in the land price by publicizing the high value of the location by the media and choosing a reasonable time for land supply. According to preliminary estimates, the industrial relocation project could increase inner-city construction land by 9.6 km^2 and generate a 13 billion rent gap in land value [49]. In addition to the land value and future land use, the industrial upgrade was guided by the municipal government after being relocated to the eastern outskirts, as were the employment and residential resettlement of previous employees.

This round of urban redevelopment brought continuous investment to the inner city and spurred the first housing boom in Chengdu from 2003 to 2008. Mainland developers and a small number of Singaporean, Taiwanese and Hong Kong investors played a significant role in promoting global urban lifestyles and modernist aesthetics that catered to a new upper class in the city. Initially, several high-end communities received investments from private developers in the southern region of the city and along the Second and Third Ring Roads. The area referred to as the European city is often recognized as the first wealthy area of the reform period. Compared to the development of high-end communities on greenfields, inner-city redevelopment was considered likely to be constrained by the land use of places occupied by work units. Newly built communities have mostly been concentrated within the waterfront area, where land was released when waterfront redevelopment ended in 1997.

These new communities have spurred a strong trend toward Western landscapes in terms of symbolic meanings (see also [37, 47, 91]). Focusing on the luxury suburban communities in Beijing, [89] defined spatial practices at the time as based on an "imagined globalization" (p. 229) by developers who want to exploit the new market for new lifestyles. Not a few developers constructed so-called "international communities" (e.g., the Athens International Community and International Community Foothills). Although not directly referred to as "International Communities," other newly built communities have also exhibited international qualities through foreign names (e.g., 17 Regent Garden; [16]). During the 2000s, these residential buildings were high-rise apartments, townhouses or detached houses. Luxury communities located on the outskirts could be deemed complete reproductions of European towns, presenting various types of so-called European architecture and public space (Fig. 4.4). Because land areas are relatively limited, inner-city communities have featured high-rise apartments of a relatively simple design, although one can still find signs of Western architecture.

Additionally, developers have simply imported images of globalization and related notions of internationalism and advanced modernity as desirable lifestyles for the new upper classes [16]. However, most of these communities have nothing to do with true international precedents. Advertisements have delivered abstract meanings of "international" related to high living standards, civilized residences and consumers who are attracted to such communities. An inner-city project called the Riverside Peak Pavilion, by a Singaporean developer, is located along the First Ring Road in the waterfront area. In promotional documents, the developer defines the main attractions of the community as follows:

Fig. 4.4 Suburban gated communities (International Community Foothills) in Chengdu. *Source* Photos taken by the author

> Luxury and hard-wearing apartments of an internationally advanced design, environments of
> a Southeast Asian style that are unique to Chengdu, high-quality Singaporean architecture,
> and services and facilities of a global standard—the community offers an international,
> modernist lifestyle. [8]

Fairview Park, the first high-end neighborhood in Chengdu located near the Second Ring Road, received investment from a Singapore capitalist in 1992 and was completed in 2000 when the city was designated by the central state as open to the world following Deng's southern tour. The developer declared that half of the community's residents were expected to be immigrants, and bilingual services were provided. Apartments in the communities were priced at 5000 Chinese yuan (approximately US $800) per square meter in the early 1990s. The catchy sales slogan "driving a Benz and living in Fairview Park" marks the transitioning lifestyles of China's new rich [12].

> What kind of lifestyle do people like in Chengdu? One can buy deep-fried dough sticks in his
> or her community or find a small restaurant at the corner of a street. However, the rich prefer
> to live in Fairview Park. It is an authentically international place, providing management
> and services just like a residential hotel. Living here is the embodiment of the real rich of
> Chengdu. [12]

As new landscapes form, residences in post-reform China have rapidly become associated with class distinctions. Chinese developers are tastemakers who create and guide housing consumer culture in China; moreover, they portray images of European architecture as iconic of middle- and upper-class urban life. Chengdu, a so-called ordinary city according to the rest of the world, has ultimately shaped its self-representation as a global city [36]. Since the city was opened to the world, this self-absorbed global image has, for the first time, produced a modernist, middle-class habitus and has driven the first wave of gentrification in Chengdu.

Apart from the newly built mode, another series of urban redevelopment plans is related to the cultural preservation and revitalization of inner-city historic places. However, the cultural revitalization of historic places in large Chinese cities has mostly adopted the commercialization mode, focusing on functional relations between economic development, urban culture, and aestheticization (see [15]). Moreover, modes of urban revitalization have involved the creative destruction of authentic places, which are replaced with modern buildings and fabricated historical structures. In four formally designated historical sites in the inner city of Chengdu, only a few key registered historic buildings remain intact following redevelopment.

A historic area featured the spatial form of *hutong* in Chengdu (Wide and Narrow Alleys, WNA), renovated after a well-known project in Shanghai called Shanghai New World (see [21, 23, 24]). The place had included historic dwellings of the Qing Dynasty and the prerevolutionary period. The area experienced commercial redevelopment during the 2000s and was designated a historic and cultural district by the municipal government in 2008. Most of the traditional buildings here were demolished and reconstructed using traditional construction techniques. Currently, the WNA represents the most notable landmark of the historical culture of Chengdu (Fig. 4.5). While such projects are designed to advance political agendas and

Fig. 4.5 Historic site (Wide and Narrow Alley) after commercial renovation. *Source* Photo taken by the author

private interests through urban reconstruction, these modes of cultural revitalization compromise cultural authenticity.

4.1.3 Toward Cosmopolitanism?

The global city making accelerated in Chinese large cities in the middle and late 2000s [19]. On the 7th Plenary Session of the 10th Beijing Municipal Committee in 2009, Beijing municipal government declared the ambition to develop Beijing as the third pole of the world, which is on a par with London and New York. This new goal was based on the *2008 Global Assessment Report of World Cities* issued by GaWC (Globalization and World Cities Research Network) [82]. The report points out that Beijing has become an ALPHA + world city together with Paris, Hong Kong, Shanghai, Singapore and Sydney. The construction goal of Beijing in the future is ALPHA ++ world city. As a result, the cultural ideas in urban development became increasingly inclusive. The image propaganda film of Beijing 2008 for Olympic Games used a series of landmark buildings and residents' activities to show the people of the world a city image that blends ancient and modern and, science, technology and humanities. In 2019, Beijing hosted the second Belt and Road International Cooperation Summit Forum. The image film at this Forum continued the keynote of that in 2008, but emphasizing more the harmony of cosmopolitanism and Orientalism and futurism and traditionalism in the city.

The promotion of Shanghai's international city image reached its peak after Shanghai won the right to host the 2010 World Expo. The blending of Eastern and

Western civilizations and the embrace of modernization had become the key image of Shanghai as an open international metropolis to the whole world [78]. In 2011, the World Chapter of the Shanghai's city image promotional film expressed explicitly the integration relationship between Shanghai and today's world cities: "Romance is not only in Paris; The energy is not just in Tokyo; Creativity isn't just in New York; Fashion, not only in London; The charm is not only in Venice. The inclusiveness and diversity has become the definition of the spirit of this city where eastern and Western civilizations converge and history and modernity merge."[2]

In this context, the new rounds of urban renewal presented pluralistic values in place-making, including not only real estate development, but also the making of transnational urbanism, cultural protection as well social development. For example, in 2001 and 2002, Beijing Municipal government released *Planning for the Protection of 25 Historical and Cultural Protected Areas in the Old City of Beijing* and *Conservation Planning of Beijing Historical and Cultural City*. Based on the planning, the land-reuse in 38% of the old Beijing city is under strict control. Until 2017, Beijing's new urban master plan clearly stipulates that *Hutong* (胡同) and *Siheyuan* (四合院) should no longer be demolished in the old city. However, residential areas identified with less historic and cultural values, which were called shanty areas, were still experiencing wholesale redevelopment.

Since 2008, Guangzhou advocated the "three 'old' (old urban areas, old industrial areas and old villages) renewal program", aiming to institutionalize the implementation of urban renewal in the city. A series of policies were released to classify the purposes and patterns of land development on the brownfields. In 2004, Shanghai Municipal Government started to strengthen the protection for historical buildings and blocks. It defines the subjects under protection in urban renewal as buildings or blocks with certain cultural, scientific and artistic value and representing the various historical periods of Shanghai [68, No. 31]. In the 13th Five-Year Plan of Shanghai, the government provided the principle of neighborhood renewal to be based on "either preservation, improvement or demolition but giving priority to preservation" [94]. Nevertheless, in both of the 2009 and 2010 work plans for urban renewal, the district government of Jingan, the most concentrated area of Shanghai's old residence, clearly put forward the idea of "high starting point", "outward oriented" and "internationalization" in the direction of neighbourhood renewal, which matched with the higher standard of Shanghai in urban modernization and globalization [66, 67].

The styles of urban spaces at this time was thus diversified. On the one hand, the large cities continued to pursue the ultra-modernist cityscape. On the other hand, the place-making could blend the signs of modernism, Westernism and Orientalism, bringing into the fever of post-modernist aesthetics. Moreover, the growth of spontaneously commercial gentrification mobilized individuals in place-making, adding even more creativity to landscape making of gentrification. These changes might have developed at a slower pace and a smaller scale in the new emerging global city of Chengdu. But they did present the world the Chinese cities with a higher level of

---

[2] From website: https://list.youku.com/albumlist/show/id_2822341.html?ascending=0.

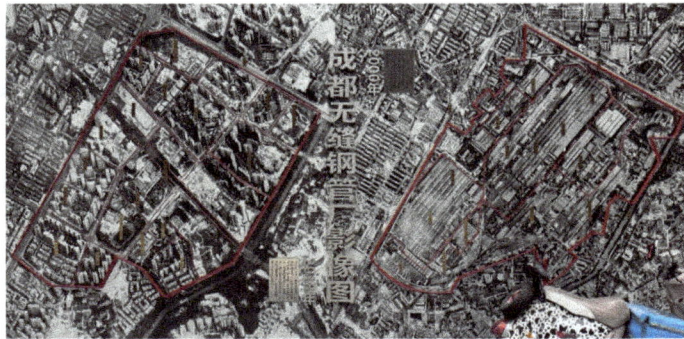

Fig. 4.6 Land-use change in the Pancheng Steel area. *Source* Photos taken by the author

modernization and cultural integration, a direction generally defined by authors as cosmopolitan urbanism [1, 18, 50].

In 2008, Chengdu was brought to the world's attention by the great earthquake that occurred in the western region of Sichuan Province, of which Chengdu is the capital city. After the earthquake, Secretary Li established a City Image Office and created a marketing corporation to promptly recover the city's image. Using the slogan, "A lucky survivor, a better Chengdu—The image of Chengdu was enhanced after the earthquake," the city's marketing scheme explained to the world that Chengdu was not at the center of the earthquake so the city could eliminate negative impressions and recover the city's tourist-friendly image.[3] Moreover, by virtue of the worldwide attention created from the earthquake, the government highlighted the efficiency of rescue teams and the government's stellar performance in spurring rehabilitation to reinforce the government's trustworthiness and to reorient the city identity as competitive and creative. In addition, the campaign made note of the investment opportunities generated through reconstruction and the resettlement of affected residents [42]. In this round of city marketing, a city video entitled "Where pandas live: Real China, Chengdu" was aired on CNN and at New York's Times Square in 2011 [93].

During this period of disaster rehabilitation, Secretary Li, in his term in office from 2009–2011, advocated for a second round of urban redevelopment. At this stage, the relocated manufacturing section and its adjacent areas mentioned earlier experienced substantial changes in function and the built environment. One of the core blocks of the industrial relocation project, where the Pancheng Steel factory was located, officially ended its industrial era in 2006 and ushered in an era of real estate development (Figs. 4.6 and 4.7). Approximately 2 km² of land were opened to the market. A number of developers at home and abroad have been stationed there since then, including three foreign-funded enterprises (Singapore Yanlord, Singapore Keppel, and Korea Lotte), four Hong Kong-funded enterprises (Sun Hung Kai, Henderson Group, Wharf, and Hong Kong Land), and six mainland enterprises (Greenland,

---

[3] Retrieved March 24, 2016, from: http://www.apexogilvy.com/caseview.aspx?id=153.

Fig. 4.7 High-end communities newly built in the Pancheng Steel area. *Source* Photos taken by the author

KWG Group, Shenzhen Greatwall, Yitai, Taihe, and General Times). The region has leapfrogged from a so-called poor region to the second highest housing price region in Chengdu, while the housing price peak is in the newly planned financial core on the south side of the city.

Yanlord Riverbay (仁恒滨河湾) in the Pancheng Steel area was developed by the Singapore developer Yanlord in 2012. The slogan "Share the transnational life with you: To the Chengdunese who are cosmopolitan and to the cosmopolitan who love Chengdu" expresses the aim of Yanlord to bring cosmopolitanism to Chengdu. The chief architecture and planner of Yanlord explained the concept of Yanlord Riverbay.

> Yanlord Riverbay will provide Chengdu with a genuine transnational lifestyle. The genuine transnational lifestyle must be backed by a transnational community atmosphere and high indoor living quality. For instance, the community clubhouse will assist community members in holding various parties; children will love to play at the exotic artificial sand beach; and adults could meet and become even business partners when they use the BBQ, yoga rooms, library and heated swimming pool. Our customers are exclusively elites with refined taste, including expatriates or those with overseas educations, successful entrepreneurs or senior managers. They have excellent educations and know exactly the lifestyle they want.

Following the eastward relocation of manufacturing, in 2007, the municipal government proposed the financial industry project of "Western Wall Street", which is located on the west side of the manufacturing area and developed alongside East Avenue. In 2009, the Chengdu Municipal Government officially issued *Opinions on Further Accelerating the Development of the Financial Industry* and formally defined the development of East Avenue as a financial street in Chengdu. With a number of preferential policies, the government aimed to attract financial institutions to settle in Chengdu. In 2010, the Chengdu government released the *Chengdu Financial Industry Development Plan (2010–2012)*, which designated the core area of a newly established high-tech zone as the financial headquarters district and the inner-city section of East Avenue (3.8 km) as the financial industry agglomeration district.

The development of the financial industry has accelerated sustainable social space replacement along East Avenue. Currently, the core East Avenue area has accumulated 152 financial institutions and 80% of the foreign banks and institutions in Chengdu. Super high-rise office buildings and five-star hotels have increased rapidly. A new generation of residential projects is also emerging, such as Times Residence developed by Hong Kong–based Wharf Holdings and the 229 m superhigh-rise apartment called New Hope D10 Tianfu.

From 2007 to 2013, Hong Kong–based Wharf Holdings alone completed 11 commercial and residential real estate projects in Chengdu. Commercial and real estate developments, branded with the "Times" prefix (e.g., Times Palace, Times First, Times Riverside, etc.) based on a Wharf Holdings product series, exhibit Wharf Holdings' determination to guide Chengdu into a new era of internationalized consumption and life. In 2014, the Executive Director of Wharf Holdings, Ms. Doreen Lee, participated in the opening ceremony for commercial real estate in International Finance Square in Chengdu (Fig. 4.8). Following the Chengdu project, Wharf Holdings will produce another four International Finance Squares in other cities of Mainland China. Using the theme "It's time," Ms. Lee described the company's reason for building in Chengdu:

Fig. 4.8 The International Finance Square in Chengdu. *Source* Photos taken by the author

Chengdu is China's third largest financial market after Beijing and Shanghai. *It's Time* for the
city to represent a world-class and coveted landmark with the opening of CDIFS. (Chengdu
International Finance Square)

"Given the high value of Chengdu's land parcels and its impressive consumption
expenditure of luxury goods, half of the Fortune 500 companies have established a
presence here; Forbes, a leading business magazine from the U.S., and CBN Weekly
from the Mainland both selected Chengdu as an emerging market supernova, and
It's Time for the people of Chengdu to be proud of themselves." Ms. Lee continued,
"The *It's Time* theme denotes a source of pride for the Chengdu people, who can
enjoy the joys and beauty of their daily lives. In addition, *It's Time* to provide the
people of Chengdu with a brand new international lifestyle and much more than just
another commercial complex for shopping" [81].

Times Riverside (时代尊邸) was the first residential estate developed by Hong
Kong–based Wharf Holdings in Chengdu. The community is located within the new
financial center and immediately adjacent to the city's commercial center. The site
previously contained traditional dwellings and the distillery of a famous liqueur in
Chengdu. In 2010, Times Riverside was completed, becoming the sole community
in the inner city to exceed 100 mu. Providing units ranging from 160 to 240 m^2 per
family, the community aimed to attract consumers with "high economic strength and
representing the new economy, new times and new nobility". The list below shows
its expectations for future customers.

Age: 35–45.

Social status: strong economic capability, high social reputation, excellent cultural taste,
trendy and healthy lifestyle.

Character: introverted and low-profile, mature, thoughtful and knowledgeable.

Values: caring life quality; pursuing healthy lifestyle; progressive in career; concerned with
the symbolic value added to consumption, namely, the consumers require living places that
are not only functional and comfortable but that also enable certain patterns of expectation
in their social meanings.

Leisure: exercises such as tennis and swimming, parties and social activities.

The last project transforming East Avenue is the eastward expansion project of
Chunxi Road Central Commercial District, which has been underway since 2010.
The project includes the construction of Taikoo Li, which is widely regarded as a
milestone in Chengdu's consumption place-making. Oakes [56, 57] argues that ideas
about the renovation of tourist places in Chinese cities reflect contradictions among
local authenticity, state ideology and transnational capital. The state ideology of
market socialism and territorialisation of cultural development has been in conflict
with and challenged by transnational capital that seeks a de-territorialised consumer
landscape. The cultural renovation in Chengdu in the early half of the 2000s, as
shown in the last subsection, more or less responds to Oakes' argument. Taikoo Li is
a typical project of cultural revitalization that illustrates the changing ideas and strategies of cultural construction in contemporary Chengdu. In contrast to the aesthetics

Fig. 4.9 Renovation process of the Daci Temple area, 2004, 2009 (above), and 2014 (below).
Source Photos taken by the author

of either traditionalism or Westernism, the local government has tended to inject local authenticity into a transnational landscape identity. The historic conservation area of the Daci Buddhist Temple includes a historical building, the Daci temple, and its surrounding area. Zoning laws established in 2005 confined construction in this historic conservation area to traditional low-rise buildings. In 2006, the district government launched a redevelopment project in the area that involved constructing archaic buildings in a Buddhist architecture style and traditional Sichuan dwellings. The plan involved removing old dwellings and relocating more than 4000 original households offsite.

In 2007, the first round of building reconstruction was completed (Fig. 4.9). However, in March 2008, the local government ordered that these buildings be demolished. The reason, which was seemingly simple and straightforward, was provided by a historical and cultural expert: "The leaders are not satisfied with the building style; it will not be a successful project, as it cannot attract commercial activities or drive real estate development in adjacent areas."[4] By the end of 2010, Swire Properties of Hong Kong and Sino-Ocean Land of Beijing bid for the land-use rights over the Daci Temple area. Over three years, the area was reconstructed as a shopping district with low-rise buildings and open streets offering high-end brands, reminiscent of

---

[4] Expert interview (O71, 2015).

Fig. 4.10 Newly built offices and condominiums in the new financial center. *Source* Photo taken by the author

the cutting-edge commercial buildings in Chengdu (Fig. 4.9). In a local newspaper article, a manager of the development company described the design concept:

> The place will provide all you need in terms of consumption. Using modern technologies, Chengdu Swire and Sino-Ocean created buildings of a traditional style. The area integrates preserved historical buildings. An international vision and creative design revive the folk culture of Chengdu. This approach respects the territorial characteristics and revives the area with new commercial activities. [9]

Other iconic spaces in the inner city of Chengdu include Lan Kwai Fong in Chengdu (developed by Lan Kwai Fong Group and completed in 2010) and the fifty-one-story Excellency Residential Apartment, Singapore Land Limited's first project in Chengdu. These new cosmopolitan spaces are all located near the inner-city financial district and waterfront areas. Unlike the residential real estate developments of the 2000s, which borrowed from European and South Asian styles, these foreign-based real estate areas are generally modernist, minimalist and spectacular (Fig. 4.10). Additionally, developers no longer promote a superficial façade of internationality for these communities; instead, developers stress symbols of success and elite nobility, particularly to financial and business professionals. In addition to promoting environmental quality, developers have constructed exclusive international kindergarten and primary schools for community members. They have built condominiums with reinforced security systems and indoor facilities, including fitness centers, cafes and clubs.

Apparently, East Main Avenue is truly a place created for elites. Metropolis Height (the name of the community) is located here; meanwhile, it is a residential area for foreigners and is close to a prosperous business center with a river view. It is exclusively for elites. It is created for the top, minor stratum of successful people, providing high-quality social networking. Here, you can speak with celebrities and tycoons, start a business with partners, party with friends and enjoy life with families and companions. Living here completely highlights your status [70].

All of these places have reshaped the landscape of the southeast corner of the inner city and ultimately became the most heavily gentrified places in the 2000s in Chengdu (including the subdistricts of Hejiangting, Shuijingfang, Niushikou and Lianxin; see Fig. 3.2c). In April 2012, after Municipal Secretary Huang Xinchu assumed office, he advocated for five city-boosting strategies that involved developing Chengdu as "the growth pole of Western China." The North Chengdu Redevelopment Program was launched and included the two oldest urban districts, Jinniu and Chenghua, and two suburban counties covering 211 km^2 and including 1.5 million residents. Currently, when taking an express bus along the elevated Second Ring Road, upon entering Chenghua District in the eastern area of the city, one can easily spot cranes and steel-framed structures lining several blocks. The program aims to create districts with "modern industries, globalized images and ecological environments" [65]. The government has presented ten action plans to finish 360 projects within the next five years. Rhetorically, the municipal secretary equated globalization with the advancement of civilization while associating urban renewal with the social development of disadvantaged groups in the city.

> The urban modernization and globalization level signifies the civilization level in a city. The development of civilization aims to advance the urban environment, to cope with the difficulties citizens experience, to cultivate modernist citizens and to build upon the comparative advantages of a city. [20]

Within the new round of the Master Plan (2011–2020), the former concept of the World Garden City, which was proposed based on the natural environment and urban form of Chengdu, was replaced with a Modernist and Global Large City concept that reflected the Chengdu government's intensified goals of economic development and ascension in achieving world-class city status. The 2013 Fortune Global Forum could be a landmark event for the integration of the city economy into the global system. During the forum, the government unveiled the city's new marketing orientation by presenting more videos about the city. Unlike the videos released in 2003, these videos focused on the newly established district, high-technological zones and spectacular imagery that drew attention to the city's active economic culture and cosmopolitan image [95].

To conclude, gentrification in China is driven by and oriented to the iterative imagination of transnational urbanism by state actors and large corporations, who are part of the Chinese elite. To date, the landscape change of Chengdu has trended toward the imitation of Western and South Asian styles, which symbolizes a better life for the new rich, and the modernist and postmodernist landscapes presenting elite lifestyles in a time of cosmopolitanism. Landscape formation shows a self-representation of globalization via spatial production by the government and developers in a Southern city. According to Ong [59], so-called modernity does not create an inverse relationship between Western and non-Western power; multiple modernities are correlated to the dynamics of global capitalism and the activities of transnational entrepreneurs in regions or countries. Cartier [5] also suggested that the creation of transitional urbanism by various agents "reveals how people create landscape and the ways in which they construct place-based meanings, identities, and expressions of broader scale ties to national and transnational arenas" (p. 1517). Based on the urban experience of Chengdu, a crucial role of the state lies in its initiation of landscape change based on the establishment of national or municipal programs [96]. Then, with governance power and discursive tools, the state could legitimate urban redevelopment by stigmatizing other places and promoting the public values of civilization and modernization. Transnational capitalists have led by inventing the styles and aesthetics of new landscapes and lifestyles within the inner city.

Nevertheless, following [14], I suggest that the role of middle-class consumers in the gentrification process remains to be seen. Based on this backdrop of cultural transformation, burgeoning middle-class consumers can be motivated to consume newly built housing and living places in their search for class distinctions in habitus. Moreover, new middle-class consumers may either gentrify newly constructed housing in the inner city or move into suburban housing, further fragmenting class distinctions. However, rather than simple manipulation and domination, the emergence of gentrification will be dependent on the extent to which urban redevelopment motivates the aspiration of distinction and subsequent consumption practices in gentrified inner-city neighborhoods, especially when compared with the simultaneous urban processes of new city building and suburbanization.

4.2 Spatial Commodification

Following the cultural ideas directing the place-making of gentrification, this section reveals the economic logic of gentrification by tracing the path to land marketization and redevelopment in neighborhood renewal projects in the inner city. It highlights the efficacy of inner-city redevelopment in accelerating the excess production in the housing market, which prompts the government to increase housing consumption in large cities in China. While it is consistent with Lees et al.'s [30] argument on the economic motivation behind state-facilitated gentrification, this discussion adds insights into the Chinese pattern of capital accumulation based on the real estate industry and the means by which the state intervenes in shaping such outcomes. Additionally, it draws attention to the significance of consumption forces in gentrification in China.

4.2.1 Two Land Markets Levels

In 1992, the World Bank released a report on housing reform in China, which included a section that offered suggestions on financial reforms for urban redevelopment [84]. The World Bank stated that inefficiencies in urban redevelopment primarily laid in project self-financing without outside funds and project costs generated from relocation plans. In addition, compensation and relocation schemes have rendered projects time-consuming and expensive, as numerous residents have relocated onsite.

Some have remained public tenants who pay the same rental fees they paid prior to redevelopment. With the exception of housing relocation, the report focused on the infrastructure provision from municipal governments and claimed that all costs will ultimately increase housing prices for new homebuyers. However, new purchasers may be reluctant to buy, as they should be presented with better choices from non-redeveloped projects.

> It appears that serviced land is underutilized by international standards. Part of the problem lies in the planner's inability to allocate an economic value to developed land and to then have the tools and flexibility to consider trade-offs between the consumption of land and other forms of capital. The costs of using land, based on compensation costs for existing and displaced users, are a poor proxy for economic value, as location or access to infrastructure is not considered. Even the value of the infrastructure associated with a site is a poor proxy because the way that infrastructure costs are recovered often bears little relation to the amount of investment actually accruing at a given location. [84, p. 67]

The report urges systematic financial reforms for urban redevelopment based on effective capital creation and financial burden-sharing. For instance, it calls for the establishment of mortgage plans that promote a cooperative mode of home financing whereby households must manage it alone. Once the housing price matches an income ratio of approximately 3:1 to 4:1, it is deemed affordable for most upper- and middle-class homebuyers. The report also calls for a complete commercialization of community facilities and criticizes urban planners in China for failing to recognize the economic value of land. The World Bank thus attempts to promote the liberation of land and housing markets insofar as differential ground rents of land can be formalized, thereby producing locational gains.

Finally, in 1994, nationwide reforms of the tax-sharing system fueled land capitalization and inspired an emerging phenomenon of land financing in Chinese cities to meet local budgets. The State Council terminated in-kind housing allocations by state-owned enterprises in 1998 and replaced them with workplace subsidies. In this context, urban redevelopment has increasingly departed from self-financing while tending toward the use of social funds and banking loans. Meanwhile, tax reforms increased burdens on local governments responsible for providing urban services. Subnational governments must share half of their revenues with the national government and are responsible for 80% of government expenditures [51]. Local governments have tended to seek land revenues by virtue of the economic responsibilities of local state actors [41, 43, 97]. However, unlike World Bank expectations, liberalization has not promoted the establishment of a perfect market that absorbs the compensation costs while preventing the elevation of housing prices by redevelopment. In contrast, the 2000s witnessed increasing densification, spatial commodification and soaring inner-city housing prices relative to those of new sites. Partly in the context of the complex land-use conditions on brownfields, the state does not necessarily minimize its influence in terms of land marketization and housing provisions in the process of urban redevelopment.

Public land ownership has generated two levels of land markets in China. These two levels are essential to understanding the interactions between state and nonstate actors in orienting land marketization. Generally, municipal land authorities act with the primary land market, either leasing land-use rights to commercial developers or allocating land to work units and village collectives at a low charge. Land transactions between commercial developers, work units and village collectives constitute the secondary market [40]. Lin and Ho [40] discovered that from 1995 to 2002, the land area illegally occupied and transacted constituted 42% of the land acquired through legal transactions. An important proportion of the illegal transactions occurred in the secondary land market. Given the price gap in obtaining state-allocated vs. commercial land, a primary pattern in the illegal transactions is to transfer state-allocated to commercial land without undergoing mandatory administrative procedures (e.g., state expropriation of collectively owned land) and government payments (see also [83]). During the same period, 86% of the land-conveyance cases in the primary market were based on informal, opaque negotiations.

State-facilitated urban redevelopment creates a formal approach to land transactions from collective to private land users. However, unlike land development on greenfields, interactions between the two land redevelopment regimes in inner-city brownfields are affected by the contextual consequences of the socialist *danwei* system in China. As a result, the work unit can de facto join the primary land market with state agencies as the supplier of land. Just prior to the land reforms, local governments and state work units, which were virtually landlords based on the socialist land system, controlled the inner-city land. For instance, in Beijing, 55% of state enterprises were located within the city's core in the early 1980s [103, p. 141], nationwide, approximately 30% of major cities were occupied by industrial enterprises and warehouses [25, p. 765]. High-level work units and local governments thus form the supply side of the land market in urban redevelopment. However, the development companies established by municipal/district governments and work units are the primary actors in land redevelopment. High-level work units are effective at protecting their land-use rights, as they are administratively subordinated to superior ministries but not directly controlled by lower levels of government. According to studies on Guangdong Province in Southern China, in cases of urban village redevelopment in peripheral areas, village collectives that have managed rural land have played a role similar to that of the work-unit systems in inner cities and can participate in the land market [26].

In most cases, land-reserve centers, local government financing platforms and state-owned development companies act within the primary land market. Put simply, local government financing platforms are government marketing departments. They are often founded temporarily for project operations and are dismantled after project termination [46]. The basic function of primary land development is land consolidation and reservation prior to land transactions. For instance, in 2015, Chenghua District in Chengdu was the site of 58 redevelopment projects, of which the government funded 31. Eight of the 31 projects were designed to promote land-use rights integration and consolidation; the others involved transportation and public facilities construction.[5] Through integration and consolidation, fragmented land-use rights previously controlled by collective and private owners have been consolidated and

---

[5] Document provided by the North Redevelopment Task Force in Chenghua District.

reserved under the Land and Resource Department. The convergence of land-use rights can thus be marketed through the land reserve center. A power shift has thus occurred from work units to state and private developers within the land-development regime.

After land consolidation, private developers bidding for land-use rights participated in design, development and construction as the main actors in the so-called secondary land market. Municipal and district governments shared the land revenues acquired through the land conveyance from the primary to the secondary land markets. However, the municipal government was likely to cede authority to the district government to overcome financial difficulties. Finally, a state-owned financing company uses the land conveyance fees to offset the demolition, compensation, and land-consolidation costs. Prior to 2004, developers with political ties could secure land-leasing fee discounts through their personal networks, while private developers with no access to government agencies were required to purchase land through official land-leasing channels, where prices could be eight to ten times higher than negotiated prices [26, p. 48]. In 2004, Document No. 71 issued by the Ministry of Land and Resources (MLR) and the Ministry of Supervision (MS) called for a strengthening of the supervision of land-leasing procedures and the redressing of illegal, negotiated transfers of land-use rights. This law required the local state to tackle or redress the incomplete or illegal land-lease approvals prior to August 31, 2004; since this date, private negotiations have been prohibited [26, 38]. Nevertheless, scholars have doubted the efficacy of central regulations on informal land leasing in the city center. While the reasons for this doubt are complex, at least three points should be noted here. Public (or the de facto public sector) land ownership was retained. For restructured SOEs, CCP party branches continue to affect company personnel systems through the control of personnel dossiers. Even for joint ventures, foreign-owned and private enterprises, which have gained tremendous discretion in their business operations, enjoy few institutionalized channels of political advantage; moreover, these companies typically resort to various SOE attachments for access to public resources such as land [60]. Furthermore, behind the black box of personal relationships (*guanxi*) and operational procedures, it is difficult to determine which land auctions are illegal or fake [101].

Nevertheless, it would be misleading to state that private developers are denied access to the primary land market. First, large state-owned enterprises and public institutes can request self-organized redevelopment with land-use and construction approval. Then, the participation can vary of primary and subsequent actors as state intervention changes strategies regarding the redevelopment market. During the 2000s, private developers charged for property acquisition, compensation and development, while local governments sponsored and managed urban redevelopment, creating special policies to attract investors and developers who could lead redevelopment projects [21]. Since 2011, however, government and legal systems have managed property acquisition and compensation. In Chengdu, state actors in the primary land market currently organize land-use rights integration and reregulate the zoning codes, while market actors in the primary land market invest in property demolition, compensation and land consolidation and construct the necessary infrastructure.

4.2.2 State Intervention and the Expanded Rent Gap

The two land market levels account for the source of state power. However, the launch of urban redevelopment contains multiple government objectives at both the central and local levels based on socioeconomic concerns. On the one hand, the Chinese government defines urban redevelopment as an economic project for "revitalizing existing state-owned assets and maintaining the appreciation of good state-owned assets" (*panhuo cunliangzichan baoliu zengzhi zichan*) (O22; see also [39]) as well as the promotion of domestic consumerism. On the other hand, the urban redevelopment program is widely popularized as "livelihood projects (*minsheng gongcheng*) for addressing housing difficulties for originally low-income residents" [73, No. 25].

Based on these multiple objectives, on the one hand, the local state tends to share economic benefits and redefine territorial power by cooperating and competing with economic actors and upper-level governance agencies [26]. On the other hand, the government is continually faced with a basic but complex issue: how can it reconcile economic growth and social questions to ensure, at the very least, a stable social transition? In particular, when the previous welfare regime was abolished, both the municipal state and *danwei*—while factoring in the socialist government, *danwei*, and society relations—were required to reorient toward new social and cultural institutions to promote state legitimacy in a market society. We then turn to the tactics and results of the state actors who intervened in the socioeconomic disputes.

Although the extent of local governments' involvement in urban renewal differs by city and project, three types of affairs generally require government participation. First, the predominant administrative power over land and housing expropriation determines state intervention in the early stage of a redevelopment project, including property acquisition and compensation, resettlement, land consolidation, land reserve and supply. He [21] identified the most important practice in Shanghai's urban redevelopment as the use of administrative means and institutional design to disentangle complex property rights issues and speed demolition. Also, local governments tend to organize in advance infrastructure and public facility construction to enhance developer investment confidence in the development area. This is called land development at the first level of the market. Early in the establishment of the land-reserve system, the primary land development is generally carried out by government agencies and land-reserve centers; that is, it is monopolized by the government. For instance, in Hefei, Anhui Province, the land-reserve center is fully responsible for the acquisition and maturation of the land stock.

Since 2011, the central government has limited the Land and Resources Departments to directly engaging in the development of the primary land market, which is delegated it to market actors who participate through a bidding process. In

Chongqing, for example, the task is the responsibility of eight state-controlled groups approved by the government. In Beijing, the government allows market participants to engage in primary land development on the premise of the overall control of the land reserve and supply plan [100]. Shanghai also follows the principle of government-led and market-oriented operation in the first level of the land market. However, the state-owned Shanghai Real Estate Group continues to be a large supporter of the government's land bank, especially for special projects such as the World Expo [32]. The Group merged with the previous Center of Land Development (established in 1996, subject to the Land and Resource Bureau), which was founded by the municipal government in 2002 [26, p. 49]. This group and its subordinate groups and companies are involved in both the primary and secondary markets. The situation in Chengdu is similar. In the early stages, the primary market was operated by the land-reserve center; later, power was delegated to state-owned and private enterprises, according to the specific nature and situation of each project.

The fragmented tenure of properties under expropriation, with a majority of public and collective assets further centralized the power of the national and munic-ipal government in formulating rules of compensation and resettlement. In 2001, the State Council (No. 305), led by former Prime Minister Zhu Rongji, limited the government's responsibility over urban redevelopment projects to supervision rather than organized demolition. Accordingly, demolition companies and devel-opers became the principal performers in housing demolition and compensation. Developers acquired approval for land use and construction from land and planning authorities (i.e., local government agents), from which they could request demoli-tion permits. In 2011, the State Council promulgated new regulations on housing expropriation and compensation. These stipulated that the main body for imple-menting demolition, relocation and compensation was the local government, and profit-making organizations were prohibited from participating.

Second, the local state is authorized to issue project-based policy incentives to attract social funds for land redevelopment. In the early stage of the paid land use reform, not a few land transfers between local governments and developers were concluded through agreements and based on non-market prices [26, 101]. Most of these developers were subordinates of state-owned enterprises or inextricably linked with administrative organs. Private developers who had no access to government agencies to gain a negotiated price for land-use rights had to gain land through official land leases, where the official price was usually eight to ten times higher than a negotiated price [26]. However, the result was that valuable land was occupied by various forces (especially large state-owned enterprises and institutions) and could not be effectively developed.

In 2004, the central government banned non-market channels for the transfer of profit-oriented land. As a result, local governments began to stimulate social invest-ments stationed in the old city through various policies. The main policy direction was to reduce or exempt land transfer fees and various taxes and to increase land-reuse flexibility [21]. In 2012, the municipal and district governments of Chengdu released policies for the North Chengdu Redevelopment Program to encourage private invest-ment. These included a full set of arrangements on land taxation, land use and construction. For example, land users were permitted to adjust land-development scales by integrating land uses in areas surrounding the redeveloped areas. Land transaction fees for the North Chengdu Redevelopment projects could be reduced to as little as 5% of the standard land price, compared with 40% for land development projects that did not benefit from special policies.[6] With the exception of land transaction fees to the government, land sale income was transferred to market actors (e.g., either government financing platforms or private developers), who should bear the cost of land consolidation as well as compensation and resettlement for original residents. The district government, which is ranked lower than the municipal government, could acquire all of the land transaction fees generated in North Chengdu Redevelopment projects without sharing them with the municipal government [10, No. 20]. The policy incentives that could be gained only by participating in state projects determined the work unit dependence on the government in land redevelopment. According to one *danwei* manager, although land redevelopment involves an asset drain for work units, the limited administrative power of state-owned enterprises, especially after market reforms, prevents work units from self-organizing redevelopment without land transactions, which essentially reflects the institutional conditions in China generated by the incomplete market transitioning of state-owned enterprises (O34). As the state continues to impose "social responsibility" on state-owned enterprises, they do not have access to the corresponding administrative networks to mobilize departmental resources and gain policy support.

> Without new investment in land, *danwei*, by independently running redevelopment, cannot balance costs. So here is the point. State-owned enterprises must often assume some social responsibilities. However, they cannot mobilize the resources, such as government powers, to fulfil these responsibilities. When the government initiates a program, it mobilizes all departments and sectors to enact favorable policies and open channels to facilitate the program, but this is not a realistic possibility for projects organized by state-owned enterprises. However, some monopolies can do this. In this case, actually the government must rely on them. For us, lower level state-owned enterprises, we must depend on the government. (O24, manager of the district government financing platform)

Finally, throughout the project's implementation, the district government and the primary land developers could urge local planning institutions to revise the regulatory plans (similar to the zoning situation) to secure profitable land-use conditions for secondary developers. Because the compensation amount for the original residents and the proportion of in situ resettlement increased, the land-use pressure on urban redevelopment was much higher than that of new city construction. In this case, local governments commonly balanced the costs of land development, consolidation and resettlement by increasing the plot ratio. This situation is particularly evident in Shenzhen and Guangzhou, where high-density urban villages are numerous. In 2018, Shenzhen adjusted the land-density zoning and plot volume–ratio standards in the *Urban Planning Standards and Guidelines*. The city cancelled the upper limit

---

[6] This policy applied to private developers or government platforms for financing who participated in the North Redevelopment Program via public land leasing. The ratio of land transaction fees to the standard price of land use increased with the time consumption of property demolition and land consolidation, ranging from 5 to 30%.

of the plot ratio of commercial service facilities land and increased the upper limit of the residential land plot ratio in all grades. This change shows the pressure on construction land in Shenzhen. Five 77-story residential towers will be built in redeveloped areas of Buxin Village and Shuiwei Village in Luohu District of Shenzhen City, with the plot ratio reaching 11. The plot ratio of the redevelopment areas of the Guangzhou Iron and Steel factories has generally been 5–8, and at least 17 residential buildings have exceeded 100 m.

In the Caojia Alley (CJA) redevelopment project in Chengdu, the primary land developer is a local government platform called Beixin Company. Beixin Company was organized by a state-owned real estate company affiliated with the Jinniu District Government and a real estate company subjected to the work unit who collectively owned properties in Caojia Alley (a provincial enterprise called Huaxi Construction Group). It is therefore affiliated with the construction office of the Jinniu District Government. According to both state actors and developers, the high costs of compensation and resettlement as well as private and public infrastructure expenditures constitute the main impetuses for rezoning (O23, O34). The considerable amount of in-kind and monetary compensation for original residents is the result of *danwei*–society relations institutionalized in the socialist period and the ideology of the Chinese government in which social stability takes precedence over other concerns. The primary investments of a redevelopment project include three components: fiscal appropriations from the national government, work-unit investments, and government and land mortgage loans to the redevelopment projects by the China Development Bank (CDB). Compensation costs for original residents constitute the most significant proportion of renewal project expenditures, while the only revenue source is land transactions with secondary investors or developers. According to local officials, the work unit (i.e., Huaxi Group) and the government each initially invested 5 million Chinese Yuan (approximately US $0.77 million) to establish Beixin Company. Then, the National Special Subsidy for the Renewal of Dilapidated Areas allocated merely 600 Chinese Yuan (approximately US $92) for each square meter of district renewal construction. As a result, compensation and construction costs were largely dependent on loans issued by the China Development Bank (CDB), which are ultimately to be repaid with the land conveyance amount (O23). Moreover, considerable levels of government debt, especially at the district level, pressured the district government to trade off the costs for each project by maximizing its profit-making abilities. In the context of this vicious circle of land speculation, according to one local official, a short-term local leader (e.g., at the district level) may benefit from the redevelopment project due to the political performance achieved when he or she is in office. However, over the long term, the government is constantly defective due to the substantial loan involved.

> The Chinese government likes to make rules for the market, but the market does not need any individuals to make rules. The government has too much conceit in thinking that it is a distributor of resources that is powerful enough to make rules. However, it may be reasonable to say, namely, that the government is the beneficiary of market intervention because individual leadership often achieves good performance and promotion due to a rise in GDP numbers. As an entity of the Chinese government, however, it is completely

damaged while marketization proceeds anyway. In the end, no one would make loans to the local government. The last resort would involve bankruptcy and rescaling. (O24, manager of the district government financing platform)

The Chengdu Municipal Government launched two types of tactics that officially created space for making zoning adjustments in redeveloped areas. The first scheme established a spatial plan to efficiently implement the North Redevelopment projects. The implementation plan combined two types of urban design and zoning adjustment plans and shortened the compilation, examination and approval procedures.[7] This type of spatial plan thus not only saves time but also assembles planners, architects, government actors, developers and residents to form cooperative relationships for mitigating conflicts among various parties and for ensuring the smooth operations of rezoning and urban design. Another noticeable set of tactics is related to land-use readjustments. The Chengdu Municipal Government issued a concession to state actors and developers participating in the North Redevelopment Program to readjust land-use rights on a larger scale (3000–6000 m^2) than in declining areas [4, No. 116]. In the case of the CJA redevelopment, the extension of land-use rights and the land redevelopment scale resulted in the inclusion of three commodity-housing projects constructed in 2002 into the redevelopment project. Homeowners occupying the three buildings later became the main activists that resisted the redevelopment. In addition, the local government and state-owned companies can change the planning conditions of the redeveloped areas provided a code balance is maintained at a larger zoning scale.[8] Some larger zoning areas include older neighborhoods positioned close to the redeveloped areas, while others may extend to a much larger scale. Under such arrangements, developers and the district government can adjust the planning conditions of the redeveloped areas while transplanting "disadvantaged" conditions, such as sanitation facilities, to the older neighborhoods nearby that may or may not be redeveloped in the future.[9]

With an implementation plan, the district government and Beixin Company have endeavored to make zoning changes that would intensify commodification and commercialization. Beixin Company prepared a land-use plan and estimated construction indicators based on economic land-development calculations. Using the land-use plan and indicators, the district government then communicated with the municipal planning department and negotiated zoning adjustment methods and planning conditions. For instance, both parties often call for high-density housing and commercial facility construction; they also often speculatively seek to provide upgraded services and favorable public facilities while reducing the proportion of

---

[7] Interview with urban planner (O1, 2014).

[8] Chengdu Municipality adapts a general rule of zoning management. Zoning codes for different land-use types are pre-established and legitimated as municipal planning regulations. Commonly, zoning codes are pre-issued to a new construction project as part of a plan with conditional enforcement. The first party of the project can only claim for the adjustment of land-use planning other than the general rule of zoning codes. However, the consolidated land-reuse policy connotes conditional power for the district government and developers to change the zoning codes of the redeveloped areas.

[9] Interview with urban planner (O20, 2015).

other types of public goods based on their own interests and expectations. According to the official informants, almost all of the redevelopment projects involving both residential and commercial land use, including the CJA and JW redevelopment projects examined in this study, have overstepped the limitations of the pre-established zoning code for the specific district. Commercial land use has also exceeded the original code levels more so than has residential land use. Taking as an example the CJA redevelopment project, to rehouse approximately 3000 households, thirteen high-rise apartments will be built on the original site. This planning condition caused the Jinniu District government to claim a plot ratio of 7 for residential land use and 11 for commercial land use, far exceeding the original zoning regulations that constrained residential use to a maximum plot ratio of 4. After a number of meetings were held between the Jinniu District Government and the Municipal Planning Institute, the residential land-use plot ratio was adjusted from 7 to 5.3, but the district government's proposal for commercial land use remains intact.[10] This situation means that the developers were able to construct higher density housing and commercial spaces than were previously allowed.

> The district government is still not satisfied with the final result (regarding plot ratios) in regards to the economic results. It is also not easy for us to ask for 7. We know the area will be far too dense, like a concrete forest or a high-rise village in the city. Therefore, we understand why the Planning Bureau rejected the proposal. The district government and the Planning Bureau have negotiated several times formally and informally and almost disagreed with each other at the last meeting. However, during the last meeting, they decided to reduce the plot ratio from 7 to 5.3. Financially speaking, this ratio is far lower than what we expected. (O24, manager of the district government financing platform)

This study does not critique whether the compensation excuses the developer for maximizing profits or whether the local government financing platform makes reasonable economic calculations. First, the study echoes arguments by Shin [69] that, because the local government depends on land financing, there is a strong tendency toward speculative land development and urbanization. Then, this study shows that in the process of urban redevelopment, the Chinese state has not only intended to spur economic growth through land marketization but has also intervened in housing improvement for low-income residents. Ideologically, the satisfaction of working-class benefits is rooted in the imperative for government legitimacy [26, 39]. Meanwhile, the Chinese state has tended to transfer the housing costs for the urban poor to land capitalization, i.e., the so-called rent gap between currently capitalized and anticipated ground rents. This mode of land redevelopment exaggerates the degree of land capitalization and spatial commodification, stimulates soaring land and housing prices and high-end inner city property development while it increases compensation and overloads the housing stock. As a result, land overproduction leads to an accumulation mode that has greatly relied on a consumer-driven economy.

---

[10] Interview with urban planner (O1, 2014) and manager (O24, 2015) of the district government platform for financing.

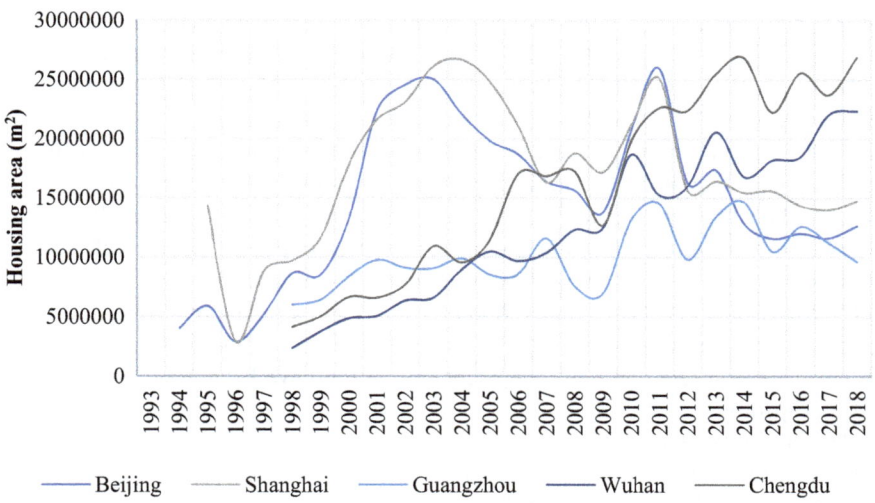

Fig. 4.11 The newly begun housing in China's large cities. *Source* City Statistic yearbooks

4.2.3 Toward a Consumer-Driven Economy?

Rule-making in land reuse ultimately exaggerates the degree of land capitalization and spatial commodification; stimulates soaring land and housing prices, high-end property development and high density in the inner city; and causes parallel increases in relocation compensation. Residential investment and housing supply continued to grow rapidly throughout the 2000s, especially before the financial crisis in 2008. In particular, Beijing and Shanghai, the national central cities, attained two peaks of annual residential construction: in the first half of the 2000s and in the early 2010s (Fig. 4.11). Since 2010, the state's control has been strengthened over the commercial housing market. As a result, the construction of commercial housing in national central cities slowed significantly after the second peak.

In contrast, provincial capital cities, such as Chengdu and Wuhan, have continued to experience even more intense fluctuations after 2010, which implies that these city economies depend on the real estate industry and its massive consumption potential. Ultimately, the real estate industry in Guangzhou has been relatively stable over these years compared with that in the other cities. This phenomenon is attributed to the much larger scale of old neighborhoods that impede the urban redevelopment process in Guangzhou.

During the 2000s, new housing construction was dominated by commodity housing developed by private developers rather than by affordable housing constructed by the state [27]. The excessive housing stock ultimately pushed the central government to introduce financial instruments to maintain the performance of the domestic housing consumer market. Since 1996, the Central Bank of China has formally allowed commercial banks to issue housing loans to individuals. Before 2001, the central government largely promoted housing consumption. During the

2000s, however, a series of financial and fiscal regulations were gradually established to control the housing bubble. In 2007, in particular, the People's Bank of China (PBC) strengthened its controls on housing speculation by claiming a minimum down payment of 40% of the housing transaction fees for a family's second house and a loan interest rate 1.1 times higher than the base rate [62, No. 359]. During the global financial crisis, the central government temporarily restored its policy incentives for housing investments. In 2008, the minimum down payment for all forms of individual housing consumption and investment was reduced to 20% of all housing transaction fees, with an interest rate 0.7 times higher than the base rate [63, No. 302]. Since the 2008 Chengdu earthquake, additional special conditions have been offered to disaster-affected districts in Sichuan Province, including Chengdu, where the loan interest rate and down payment proportion have been reduced to 0.6 and 10%, respectively.

Housing investment policy incentives brought about soaring housing prices in 2009 that motivated the central government to initiate a second round of macro-regulation in 2010. In 2010 and 2011, the most influential policies to date were released to curb housing speculation. For instance, the down payment ratio for second home purchases was increased to 60% of the housing price. Moreover, households avoided purchasing a third house. Other systematic factors (e.g., local *hukou*) can restrict individual behavior in real estate investment [72, No. 1]. The second round of housing speculation regulations, however, gradually faded following the cooling of the consumer housing market and the accumulated housing inventory. After advocating for nationwide urban renewal in 2013, Premier Keqiang Li implemented strategies to restore the housing market and maintain domestic consumption. The new policies tended to strengthen regulatory flexibility by differentiating and decentralizing decision-making among regions and cities. For example, as it has generally deregulated second home purchases (e.g., from 60% in 2011 to 40% in 2015 and 30% in 2016), the central government has also decentralized regulations in larger cities, especially in Beijing, Shanghai, Guangzhou and Shenzhen [52, No. 128, 61]. In Chengdu, for example, before 2016, the ratio of down payments for second home purchases was reduced to as low as 35% in the main urban districts and to 30% in other districts and counties [61]. Moreover, for individuals with public funds and good credit (e.g., those without housing debt), the down payment ratio for second home purchases could be as low as 20% [64, No. 128]. Periodically, speculative land development has been combined with promotions for housing consumption and investment.

These government interventions have resulted in urban development being dependent upon the consumer economy. In this type of urban development, an increase in investment and supply is always accompanied by an increase in consumption. Figure 4.12 shows five representative cities chosen to demonstrate the changes in China's housing market from the late 1990s to 2018. In the period 2000–2007, the five cities consistently showed a positive correlation between housing investment and housing sales area ($R = 0.455^{**}$). When the whole period from 1998–2018 is analyzed, we still find a positive correlation between the newly started housing area and the housing sales area (0.428^{**}; Fig. 4.13).

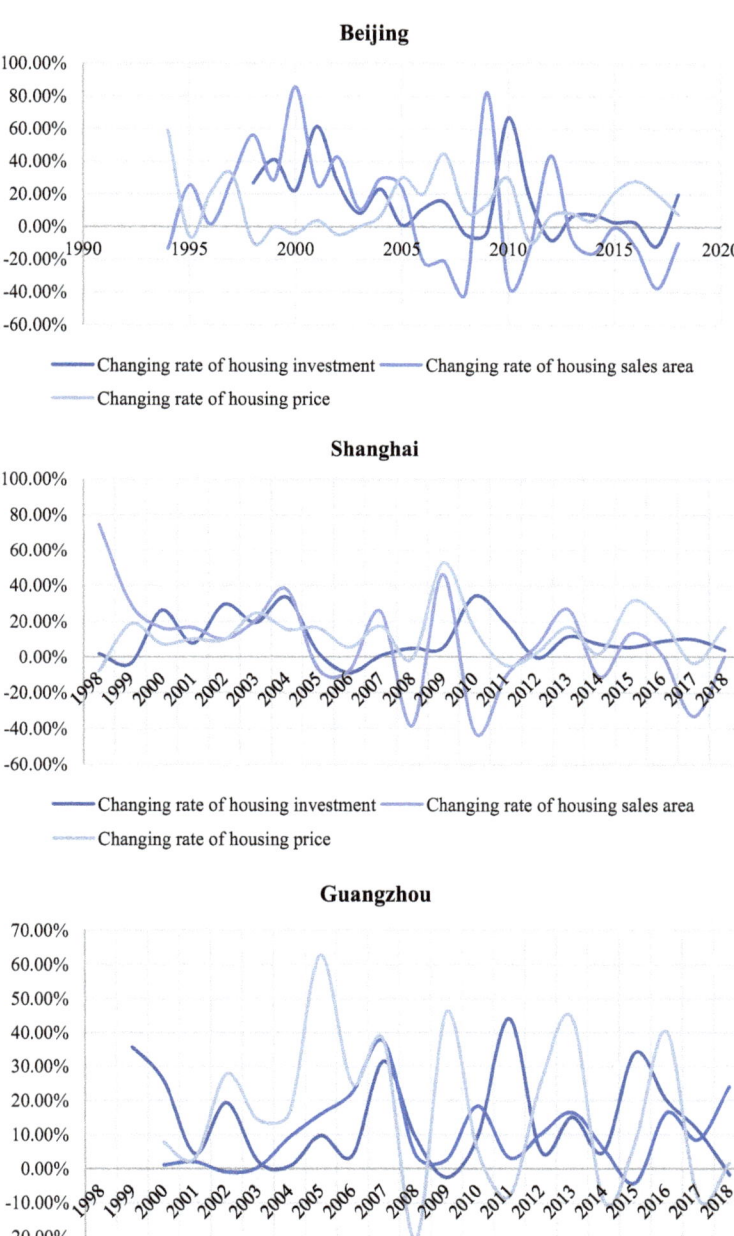

Fig. 4.12 Changing housing market in large Chinese cities

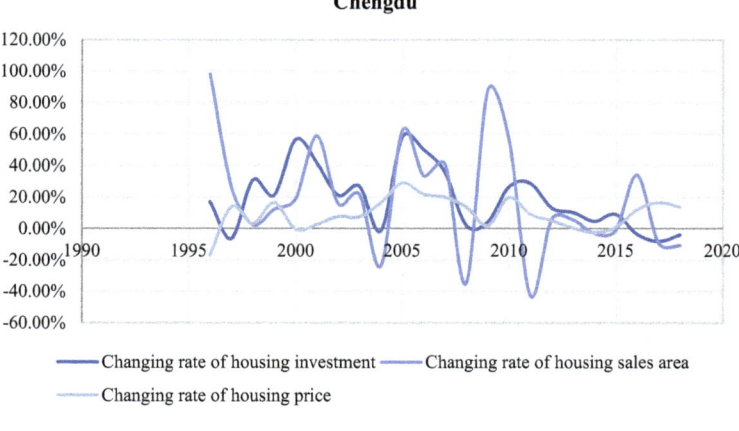

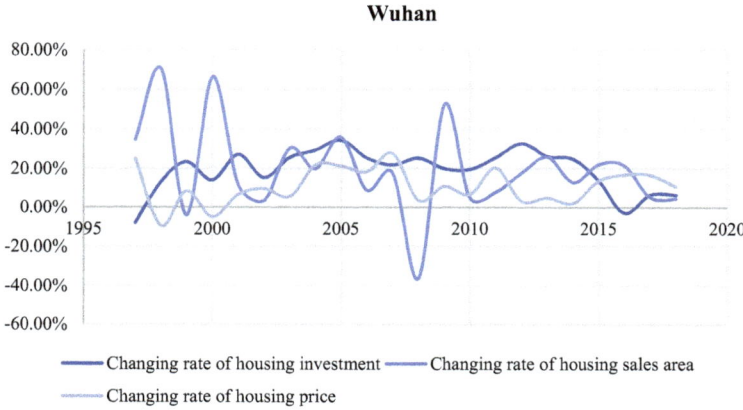

Fig. 4.12 (continued)

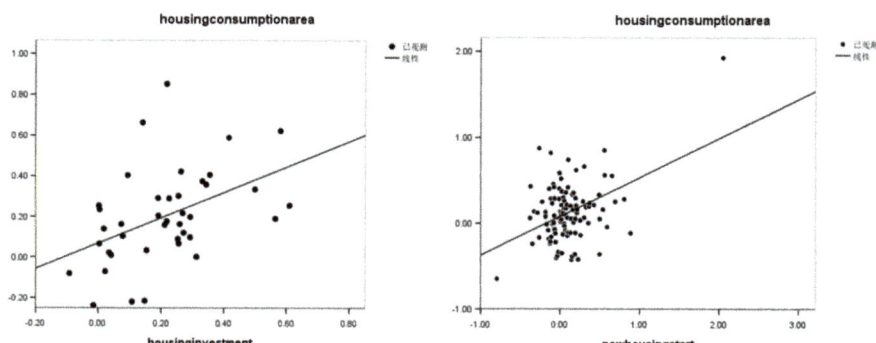

Fig. 4.13 Correlation between housing consumption and investment; housing consumption and newly begun housing

On the one hand, this positive correlation adds evidence to the effects of the government and developers' stimulating consumption through financial means; on the other hand, it reflects consumers' confidence in the market and their massive consumption potential. Among these cities, the synchronization of changes is the most obvious in the housing supply and consumption in Shanghai and Chengdu, suggesting the activity of real estate industries in the city economy. Moreover, in many years, the growth of housing prices and sales areas in Shanghai and Chengdu attained simultaneous peaks, which reflects a higher expectation of residents for the city's housing market and a relatively speculative attitude in housing purchases. Since 2008, the real estate market in China has experienced a series of changes, such as the financial crisis of 2008 and the release of national ordinances to curb the excessive rise in house prices in 2010. The average growth rates of the housing investment, prices and sales have all slowed since 2010.

In examining the production and consumption relationship, Cartier [7] argued that the urban economy in China is characterized by hybrid conditions. The urban economy relies on *excess production* for growth; meanwhile, "the large surpluses perpetuate the problem of how to absorb them" (p. 61). This excess production is most evident in the construction of prior infrastructure, superfluous commercial districts, and unsold residential spaces. Based on these hybrid conditions, Cartier mainly argued [5–7] that excess production necessitates the encouragement of "excess" consumption. Moreover, the imbalance in societal transformation and overproduction in the built environment is a basic motive for the state to circulate social learning, such as the civilization program, to either cultivate consumer citizenry or control resistance to rapid urban transformation.

> In the process of seeking to rebalance the economy from production- to consumption-driven, the production of consumer space in the built environment only sets the stage. Local urban authorities throughout China appear to widely accept the developers' mantra of "build it and they will come." Through the 1980s, most households continued to prioritize savings. After slow growth in the 1990s, transfers of economic surplus to households through wages, supplemented by grey income, steadily increased in the 2000s until the Global Financial Crisis of 2007–2009. [7, p. 61]

Based on the land-redevelopment process, I argue that state-facilitated gentrification contributes to speculative surplus accumulation through overproduction and the ensuing encouragement of spatial consumption. In part, the result is conditioned by the specific social and institutional settings of brownfield sites. According to the previous analysis, the outcome is also reinforced by the potential of inner cities to support cultural innovation, thereby unearthing a new consumer market. In this case, gentrification becomes dependent on the performance of a consumer-driven economy in the city. If the government constantly issues policies to maintain efficient consumption and speculation, gentrification is more likely to emerge. This revelation calls for a rethinking of the social process linking land reinvestment and gentrification in Chinese cities, in particular, by emphasizing the consumption side of gentrification.

References

1. Beck U (2004) Cosmopolitical realism: on the distinction between cosmopolitanism in philosophy and the social sciences. Global Netw 4(2):131–156
2. Bray D (2005) Social space and governance in urban China: the Danwei system from origins to reform. Stanford University Press, Stanford, CA
3. Bray D (2006) Building "community": new strategies of governance in urban China. Econ Soc 35(4):530–549
4. Bureau of Land and Resource of Chengdu (2012) Chengdu beigai tudi liyong shishi xize (shixing) (regulations on the implementation of the North City Redevelopment programme in Chengdu (trial)), No. 116. Retrieved 11 July 2015, from: http://www.cdlr.gov.cn/detail.aspx? id=37808
5. Cartier C (2002) Transnational urbanism in the reform-era Chinese city: landscapes from Shenzhen. Urban Studies 39(9):1513–1532
6. Cartier C (2009) Production/consumption and the Chinese city/region: cultural political economy and the feminist diamond ring. Urban Geogr 30(4):368–390
7. Cartier C (2016) Governmentality and the urban economy: consumption, excess, and the 'civilised city' in China. In: Bray D, Jeffreys E (eds) New mentalities of government in China. Routledge, New York, pp 56–73
8. Chengdu Commercial Daily (2012) Women Xiang Chengdu tigong gaoduan zhuzhai (we provide high-end residence to Chengdu). Retrieved 23 July 2015, from: http://e.chengdu.cn/ html/2012-08/02/content_339256.htm
9. Chengdu Evening Post (2014) Chongshu chengshi zhongxin: Chengdu Yuanyang Taiguli jiuyue kaiye (rebuilding the city centre: sino-ocean Taikoo Li Chengdu opens in September). Retrieved 23 July 2015, from: http://www.cdwb.com.cn/html/2014-01/24/content_1992235. htm
10. Chengdu Municipal Government (2012) Guanyu jinyibu cujin beicheng gaizao de yijian (opinions on promoting north city redevelopment) No. 20. Retrieved 4 Jan 2013, from: http://wwww.chengdu.gov.cn/wenjian/detail.jsp?id=1dMaHLHiQ3vex4ndT9Di&Cla ssID=07030202090104
11. China Business News (2016) Chengdu litui laojiu yuanluo gaizao, pojie chengshi neibu eryuan jiegou (Chengdu promotes urban renewal of declining areas to deal with the duality of urban spaces). Retrieved 29 June 2016, from: http://dianzibao.cb.com.cn/html/2016-02/01/node_18. htm
12. Da A (2009) Huaxinguoji, Zaici qiaodong xin jinxiu quanceng (Huaxin international: Shaking the new Fairview Park area). West China Video Telev Post 8:104–105
13. Dan CL, Wang Q (2002) Comparison of land use planning between Western countries and China. China Land Sci 16(1):43–46 (in Chinese)
14. Davis D (2005) Urban consumer culture. China Q 183:692–709
15. Deng Q (2005) Report on Chengdu urban development, 2002–2004. Sichuan Population Press, Chengdu
16. Feng G (2006) Fangdican guanggao: Yige shidai bianqian de zhenshi wenben (real estate commercials: an authentic text of the transitional period). Dissertation, Sichuan University
17. Gaubatz P (1999) Understanding Chinese urban form: contexts for interpreting continuity and change. Built Environ 24(4):251–270
18. Glick Schiller N (2012) Situating identities: towards an identities studies without binaries of difference. Identities 19(4):520–532
19. Gu C, Wei YD, Cook IG (2015) Planning Beijing: socialist city, transitional city, and global city. Urban Geogr 36(6):905–926
20. Guangmin Daily (2015) Chengdu: Chuangjian wenming chengshi meiyou juhao (Chengdu continues in constructing Civilised City). Retrieved 14 Mar 2016, from: http://www.gmw.cn/ 01gmrb/2005-11/11/content_329768.htm
21. He S (2007) State-sponsored gentrification under market transition: the case of Shanghai. Urban Aff Rev 43(2):171–198

22. He S (2012) Two waves of gentrification and emerging rights issues in Guangzhou, China. Environ Plan A 44:2817–2833
23. He S, Wu F (2005) Property-led redevelopment in post-reform China: a case study of Xintiandi redevelopment project in Shanghai. J Urban Aff 27(1):1–23
24. He S, Wu F (2007) Socio-spatial impacts of property-led redevelopment on China's urban neighborhoods. Cities 24(3):194–208
25. Ho SPS, Lin GCS (2004) Non-agricultural land use in post-reform China. China Q 179:758–781
26. Hsing Y (2010) The great urban transformation: politics of land and property in China. Oxford University Press, New York
27. Huang Y (2012) Low-income housing in Chinese cities: policies and practices. China Q 212:941–964
28. Kai Y (1993) Zhongguo xiaokang zhuzhai tongyong tixi. (The standard of housing building for the middle-income family in China). Hous Sci 9:3–6
29. Knapp RG (1999) China's living houses: folk beliefs, symbols, and household ornamentation. University of Hawaii Press, Honolulu
30. Lees L, Shin HB, López-Morales E (2016) Planetary gentrification. Polity Press, Cambridge, Malden, MA
31. Ley D (1996) The new middle class and the remaking of the central city. Oxford University Press, Oxford
32. Li Y (2010) Tudi yiji kaifa moshi tantao—Shanghai dican jituan kaocha jilu. Discussion on the primary land development model—investigation record of Shanghai Real Estate Group). The Second Beijing Youth Land Academic Exchange Conference. Beijing
33. Li Y, Yang JW (2014) Cong Chengdu jiucheng gaizao lishi yu zhengce huishu kan zizhi gaizao moshi de xingcheng. (The formation of the self-governance mode of redevelopment in Chengdu: a review of policies and modes). J Chengdu Inst Publ Adm 3:22–25
34. Liang SC (1954) The characters of Chinese architectures. J Archit 1(2):29–36
35. Liang SY (2014) Remaking China's great cities: space and culture in urban housing, renewal, and expansion. Routledge, New York
36. Liang T (2008) The globalised city: social background and self-representation of city promotional videos of Chengdu (1999–2006). Dissertation, Hong Kong Baptist University
37. Lin GCS (2007) Chinese urbanism in question: State, society, and the reproduction of urban spaces. Urban Geogr 28(1):7–29
38. Lin GCS (2009) Developing China: land, politics, and social conditions. Routledge, London
39. Lin GCS (2015) The redevelopment of China's construction land: Practising land property rights in cities through renewals. China Q 224:865–887
40. Lin GCS, Ho SPS (2005) The state, land system, and land development processes in contemporary China. Ann Assoc Am Geogr 95(2):411–436
41. Lin GCS, Yi F (2011) Urbanisation of capital or capitalisation on urban land? Land development and local public finance in urbanising China. Urban Geogr 32:50–79
42. Liu L (2009) Chengdu yi nyuqiyejia yin chaiqian louding zifen (an entrepreneur in Chengdu burns due to housing demolition). New Hunan Newspaper. Retrieved 13 Aug 2015, from: http://www.boxun.com/news/gb/china/2009/11/200911262243.shtml
43. Liu T, Lin GCS (2014) New geography of land commodification in Chinese cities: uneven landscape of urban land development under market reforms and globalisation. Appl Geogr 51:118–130
44. Liu Y, Zhou Y (2021) Territory spatial planning and national governance system in China. Land Use Policy 102:105288
45. Lu J, Luo P, Zhang J (2002) Zhongguo xiandai chengshi zhuzhai. (modern urban housing in China, 1840–2000). Qinghua University, Beijing
46. Lu Y, Sun T (2013) Local government financing platforms in China: a fortune or misfortune? International Monetary Fund
47. Ma LJC (2002) Urban transformation in China, 1949–2000: a review and research agenda. Environ Plan A 34:1545–1569

48. Ma LJC, Wu F (2005) Restructuring the Chinese city: diverse processes and reconstituted spaces. In: Ma LJC, Wu F (eds) Restructuring the Chinese City. Routledge, London, UK, pp 1–20
49. Ma RQ (2004). Fenghuang shi zenyang niepan de—Chengdu dongjiao gongyequ jiegou tiaozheng shilu. How did the phoenix become nirvana? memoir of structure adjustment in East Suburban Industrial Zone of Chengdu, CO. Company 5:4
50. McFarlane C (2011) Assemblage and critical urbanism. City 15(2):204–224
51. Man JY (2010) Local public finance in China: an overview. In Man JY, Hong Y-H (eds) China's local public finance in transition. Lincoln Institute of Land Policy. Cambridge, MA, pp 3–20
52. Ministry of Housing and Urban-Rural Development, People's Bank of China and Ministry of Finance (2015) Guanyu tiaozheng zhufang gongjijin geren zhufang daikuan shoufu bili de tongzhi (notification on adjusting the minimum down payment ratio for the purchase of housing units with individual housing provident fund loans), No. 128. Retrieved 24 Apr 2016, from
53. National People's Congress (2018) Zhonghua renmin gongheguo xianfa xiuzhengan (constitution). Retrieved 12 Feb 2021, from: http://www.npc.gov.cn/npc/c505/201803/e87e5cd7c1ce46ef866f4ec8e2d709ea.shtml
54. National People's Congress (2019). Zhonghua renmin gongheguo tudiguanlifa (Law of Land Administration). Retrieved 12 Feb 2021, from: http://www.npc.gov.cn/npc/c30834/201909/d1e6c1a1eec345eba23796c6e8473347.shtml
55. Ni P, Chen J (2010) Strategies of Chengdu's international City marketing. Social Science Academic Press, Beijing
56. Oakes T (1998) Tourism and modernity in China. Routledge, London
57. Oakes T (1999) Bathing in the far village: globalisation, transnational capital, and the cultural politics of modernity in China. Posit East Asia Cult Critiq 7(2):31–66
58. Ong A (1991) The gender and labor politics of postmodernity. Annu Rev Anthropol 20(1):279–309
59. Ong A (1997) Chinese modernities: narratives of nation and of capitalism. In: Ungrounded empires: the cultural politics of modern Chinese transnationalism, pp 171–202
60. Pearson MM (1997) China's new business elite: the political consequences of economic reform. University of California Press
61. People's Bank of China and China Banking Regulatory Commission. (2016) Guanyu tiaozheng geren zhufang daikuan zhengce youguan wenti de tongzhi (notification on adjusting individual housing loan policies). Retrieved 24 Apr 2016, from: http://www.pbc.gov.cn/gou tongjiaoliu/113456/113469/3014377/index.html
62. People's Bank of China and China Banking Regulatory Commission (2007) Guanyu jiaqiang shangyexing fangdichan xindai guanli de tongzhi (notification on strengthening the administration of the credit loans to commercial real estates), No. 359. Retrieved 24 Apr 2016, from: http://www.fotic.com.cn/tabid/117/InfoID/754/frtid/102/Default.aspx
63. People's Bank of China (2008) Guanyu kuoda shangyexing geren zhufang daikuan lilv xiafu fudu youguan wenti de tongzhi (notification on reducing the interest rate for individual housing loans), No. 302. Retrieved 24 Apr 2016, from: http://www.pbc.gov.cn/bangongting/135485/135495/135497/818711/index.html
64. People's Bank of China (2015) Guanyu tiaozhen zhufang gongjijin geren zhufang daikuan zuidi shoufukuan bili de tongzhi (notification on adjusting the down payment for individual housing purchase using housing loans), No. 128. Retrieved 24 Apr 2016, from: http://www.mof.gov.cn/zhengwuxinxi/zhengcefabu/201509/t20150901_1444424.htm
65. People's Daily (2012) Chengdu maixiang xibu hexin zengzhangji (Chengdu towards the main growth pole of the west China). Retrieved 5 Sept 2015, from: http://paper.people.com.cn/rmrb/html/2012-03/01/nw.D110000renmrb_20120301_9-01.htm
66. Shanghai Jingan District Government (2009) Guidance opinions on further accelerating the regeneration of old districts in Jingan District (Guanyu jinyibu jiakuai jinganqu jiuqugaizao de zhidao yijian). Retrieved 23 Feb 2022, from: https://www.jingan.gov.cn//govxxgk/JA1/2009-05-04/7cc607fe-93de-48ba-8161-43e31913a2e2.html

67. Shanghai Jingan District Government (2010) Opinions on further accelerating the regeneration of old districts in Jingan District (Guanyu jinyibu jiakuai jinganqu jiuqugaizao de shishi yijian). Retrieved 23 Feb 2022, from: https://www.jingan.gov.cn/govxxgk/JA1/2010-03-02/2a609d4d-0130-45ec-8b5a-62cd6547cb48.html

68. Shanghai Municipal Government (2004) Notice on further strengthening the protection of historical and cultural areas and excellent historical buildings in Shanghai (Guangyu jinyibu jiaqiang benshi lishiwenhua fengmaoqu he youxiu lishijianzhu de tongzhi). Retrieved 23 Feb 2022, from: https://law.lawtime.cn/d361256366350.html

69. Shin HB (2016) Economic transition and speculative urbanisation in China: gentrification versus dispossession. Urban Stud 53(3):471–489

70. Shouhu Focal Point (2014) Huazhi, Duhuihuating: Yu xibu huaerjie yuliang shijieji caifu mengxiang (Chinese Estates Group, metropolis height: dreaming the world affluence in the wall street in the Western China). Retrieved 26 Jan 2016, from: http://news.focus.cn/cd/2014-06-18/5163881.html

71. Song Y (2016) Study on realization mechanism of land planning power in China. Zhejiang University (in Chinese)

72. State Council of the People's Republic of China (2011) Guanyu Jinyibu zuohao fangdichan shichang tiaokong gongzuo youguan wenti de tongzhi (notification on strengthening the regulations of the real estate market), No. 1. Retrieved 16 July 2015, from: http://www.gov.cn/zwgk/2011-01/27/content_1793578.htm

73. State Council of the People's Republic of China (2013) Guanyu jiakuai penghuqu gaizao gongzuo de yijian (opinions on accelerating the shantytown renewal), No. 25. Retrieved 13 July 2015, from: http://www.gov.cn/zwgk/2013-07/12/content_2445808.htm

74. Sun P (2010) Report on Chengdu urban development, 2007–2009. Sichuan Population Press, Chengdu

75. Tao W, Wang R (1999) Chengdu wushi nian, 1949–1999 (fifty years of Chengdu, 1949–1999). China Statistic Press, Beijing

76. Tian L (2007) Regulatory detailed planning in China: Dilemma and solution. City Plann Rev 31(1):16–20 (in Chinese)

77. Tianfu Morning Post (2005) 2005 Chengdu: Yihan (unfortunate Chengdu, 2005). Retrieved 26 Jan 2016, from: http://morning.scol.com.cn/2005/12/30/20051230704024659442.htm

78. Wang H, Xiaokaiti M, Zhou Y, Yang Y, Liu Y, Zhao R (2012) Mega-events and city branding: a case study of Shanghai world expo 2010. J US-China Publ Admin 9(11):1283–1293

79. Wang LN (2006) Zhongguo chengshi tudi chanquan zhidu yanjiu (the land system in urban China). Social Sciences Academic Press (in Chinese)

80. Wang Z (2007) Chengbian: Yige zhongguo chuantong chengshi de xibao zaizhao (Urban change-The renovation of a historic city). People's Publishing, Beijing

81. Wharf Holdings Limited (2014) Wharf unveils its Chengdu international finance square setting a New City landmark in China West bringing hundreds of world-renowned brands to China's "new first-tier" city. Retrieved 26 Jan 2016, from: http://www.wharfholdings.com/file/140114%20CDIFS%20Opening%20Eng%20P%20FW.pdf

82. Wang H (2011) A study on the creation of "creative city" brand in Beijing (Beijing shuzhao chengshi chuangyizhidu pinpai yanjiu). In: 2011 City internationalization forum--metropolitan governance in the process of globalization

83. Wong KK, Zhao XB (1999) The influence of bureaucratic behaviour on land apportionment in China: the informal process. Environment and Planning C 17:113–126

84. World Bank (1992) China: implementation options for urban housing reform. World Bank, Washington

85. Wortzel LM (1987) Class in China: stratification in a classless society. Greenwood Press, New York

86. Wu F (1996) Changes in the structure of public housing provision in urban China. Urban Stud 33(9):1601–1627

87. Wu F (2000) The global and local dimensions of place-making: remaking Shanghai as a world city. Urban Stud 37(8):1359–1377

88. Wu F (2000) Place promotion in Shanghai, PRC. Cities 17(5):349–361
89. Wu F (2004) Transplanting cityscapes: the use of imagined globalization in housing commodification in Beijing. Area 36(3):227–234
90. Wu F (2015) Planning for growth: urban and regional planning in China. Routledge, New York, London
91. Wu F (ed) (2007) China's emerging cities: the making of new urbanism. Routledge, New York
92. Xia J, Li T, Jiang ZJ, Zhou L (2021) Deviations and mechanisms in the allocation of construction land quotas from the perspective of national governance. Sci China Land 35(6):20–30 (in Chinese)
93. Xu R (2011) Chengdu chengshi xingxiang xuanchuanpian denglu niuyu shidai guangchang (Chengdu video shows in time square). Xinhua Net. Retrieved 14 Mar 2016, from: http://news.xinhuanet.com/world/2011-08/02/c_121760963.htm
94. Xu X (2019) From demolition, renovation and retention to retention, renovation and demolition—renovation of old districts and protection of historical features in Shanghai since the new century (Cong chai, gai, liu dao liu, gai, chai—xinshiji yilai Shanghai jiuqu gaizao yu chengshi lishifengmao baohu). Party Hist Party Build Shanghai 2:5
95. Yan XL (2015). Chengdu chengshi xingxiang xuanchuanpian zhongde chengshi wenhua jiangou (cultural construction in the advertising videos for the city image of Chengdu). Qingnian Jizhe (Youth Journalist). Retrieved 14 Mar 2016, from: http://qnjz.dzwww.com/gdst/201501/t20150130_11821274.htm
96. Yang M (1997) Mass media and transnational subjectivity in Shanghai: notes on (re)cosmopolitanism in a Chinese metropolis. In: Ong A, Nonini D (eds) Ungrounded empires: the cultural politics of Chinese transnationalism. Routledge, London, pp 287–319
97. Yeh AGO, Wu F (1996) The new land development process and urban development in Chinese cities. Int J Urban Reg Res 20(2):330–353
98. Zhang J (2003) Chengdushi dongjiao gongye jiegou tiaozheng guihua (structural adjustment planning of east suburban industrial zone of Chengdu city). Sichuan Arch 23:36–37
99. Zhang SH (2012) The problems and countermeasures of the system of forced house removal—comment on the regulation of house expropriation and compensation on state-owned land. Law Rev 30(3):102–107 (in Chinses)
100. Zhang YR (2014) Tudi yijikaifa liyifenpei ji xietiaojizhi yanjiu (study on benefit distribution and coordination mechanism of primary land development). Chongqing University
101. Zhang Y, Fang K (2004) Is history repeating itself?: from urban renewal in the United States to inner city redevelopment in China. J Plan Educ Res 23(3):286–298
102. Zhao G (1991) 2000 nian de zhuzhai. (Housing in 2000). China Architecture and Building Press, Beijing
103. Zhou Y, Meng Y (2000) Beijing de jiaoquhua jiqi duece (Suburbanisation and Policies in Beijing). Science Publishing, Beijing
104. Zhu JM, Liu X, Tian L (2007) The relationship between land use planning and individual land rights—the impact of law of property rights on urban planning. Urban Plann Forum 4:56–64 (in Chinese)
105. Zi D, Huang J (2006) Guanyu Chengdu jiucheng gaizao de sikao. China Economic Times. Retrieved 15 July 2015, from: http://www.sc.xinhuanet.com/scls/content/2006-08/10/content_7746420.htm

Chapter 5
Convergence

This chapter turns to the consumption force in gentrification. As described in Chap. 1, a status quo of class transformation in the current cities is captured in the ongoing formation/fragmentation of class-based habitus and identities. Moreover, consumption practices become instrumental as individuals construct and declare class distinctions and identities in the market society of post-Mao China. In this context, this chapter uncovers the attributes and motives of gentrifiers to move into the newly built areas. Unlike the urban pioneers who championed inner-city urbanism in the post-1960s in Western society, this study shows how the sociocultural temperament of Chinese gentrifiers relates to the new inner-city urbanism in China and contains sociopolitical meanings.

5.1 Socioeconomic Attributes of Gentrifiers

Two subdistricts, the most (Lianxin) and the least (Hehuachi) gentrified, were selected to analyze the socioeconomic profile and other attributes of gentrifiers. Lianxin subdistrict contains high-rise riverside residences built in the early 2000s and the representative works of the newest generation of residences in Chengdu. The gentrified areas are mainly located in the Jiuyanqiao community under the jurisdiction of Lianxin. The community is located on the riverside and at the east side of the nearby historic site of Jiuyan Bridge. Starting in 1958, the area was developed as a factory *danwei* residence.

Table 5.1 presents the socioeconomic attributes of Lianxin and Hehuachi subdistricts in both 2000 and 2010. From the perspectives of education and the hukou population, the two neighborhoods presented comparable conditions in 2000, with the secondary educated population being the largest group (approximately 61% of the total population over 6 years old) in each subdistrict and with rural–urban migrants amounting to 20% of the entire population. Nevertheless, by 2010, 28.06%

© Springer Nature Singapore Pte Ltd. 2022
Q. Yang, *Gentrification in Chinese Cities*, Urban Sustainability,
https://doi.org/10.1007/978-981-19-2286-2_5

Table 5.1 Socioeconomic profiles of the most and least gentrified subdistricts

	Lianxin (the most gentrified)		Hehuachi (the least gentrified)	
	2000	2010	2000	2010
Education				
Primary education	20.60	7.03	17.79	16.45
Secondary education	61.96	41.04	61.00	63.93
College degree	8.28	20.18	11.75	10.83
University and graduate degree	4.08	28.06	5.50	6.71
Household registration status				
Rural hukou population	20.06	17.88	20.15	35.68
Migrants	48.48	39.82	49.11	53.18
Industry				
Manufacturing workers	34.39	18.58	14.00	5.05
Retail trade workers	13.91	17.74	14.38	27.21
Personal service workers	8.96	10.51	24.7	32.89
Producer service workers	6.96	14.53	4.17	6.03
FIRE workers	4.40	9.63	2.02	4.00
Business services workers	2.55	4.89	2.15	2.03
Social service and public administration	12.52	15.06	11.45	6.77
Public administration workers	4.25	4.66	3.89	1.71
Social service workers	8.27	10.4	7.57	5.07
Occupation				
Managers	4.80	8.87	3.26	0.36
Enterprise directors	3.03	8.68	2.21	0.30
Managers in public institutions	1.31	0.11	0.05	0.02
Managers in government sectors	0.46	0.08	1.00	0.04
Professionals	17.77	29.63	16.59	8.92
Financial business workers	0.54	4.20	0.48	1.04
Clerk and related workers	12.04	13.88	14.00	16.42
Commerce, service trade personnel	35.70	34.90	43.28	62.76
Production workers	28.90	12.39	21.33	13.60
Tenure type				
Self-built homeowners	2.83	1.16	0.67	0.4
Public tenants	26.44	3.06	19.87	2.91
Private tenants	3.73	13.29	3.82	32.42
Owner-occupants of commercial apartments	5.85	49.91	3.82	16.04
Subsidized owners	26.74	25.05	51.57	34.87

(continued)

Table 5.1   (continued)

	Lianxin (the most gentrified)		Hehuachi (the least gentrified)	
	2000	2010	2000	2010
Owner-occupants of price-controlled housing	17.15	4.31	9.43	2.74
Housing expenditure				
Average monthly rent	130.27	1050.00	215.73	721.63
Locational character				
Middle school		6.50		3.50
Distance to financial and business center		2.50		4.00
Distance to historical and cultural sites		2.50		4.00
Distance to traditional commercial and business center		4.00		4.50
Housing condition				
Households in housing built after 1990	55.23	82.87	36.53	47.40
Average floor space of building (per population)	19.47	30.41	18.01	23.58
Households with independent bath	69.32	89.49	76.10	86.64
Households in building over than 7 stories	40.80	67.79	19.37	41.14
Households in reinforced concrete building	49.49	71.70	15.75	43.44

Note The numbers in this table refer to the "percentage of total population." *Source* Computed based on the population censuses of Chengdu and spatial planning materials provided by the Chengdu Urban Planning and Research Institute

of Lianxin's population had a university degree, while Hehuachi continued to principally accommodate a population with a secondary education (63.93%). Meanwhile, rural–urban migrants decreased to 17.88% in Lianxin, compared with 35.68% in Hehuachi.

From 2000 to 2010, the employed population nearly doubled in Lianxin (15,440–26,160), the most gentrified subdistrict. When they were classified by industry, these workers were evenly distributed between the personal, producer, and social and public service sectors, with the lowest proportion being from the personal service industries (10.51%). The labor force working in producer services accounted for 14.53% of all employees, while the share of those employed in the financial, insurance and real estate (FIRE) industries reached 9.63%. In absolute numbers, the communities in Lianxin Subdistrict attracted an additional 2726 producer service workers over ten years, of which 1840 were engaged in the FIRE industries. Unlike the uptick in producer service workers, social and public service workers decreased in a majority of the inner-city subdistricts. In particular, Lianxin witnessed minimal growth (2.54%) of the social group from 2000 to 2010, while it fell by 4.68% in Hehuachi. In addition, although in decline, Lianxin still has a group of manufacturing workers representing 18.58% of the employed population in 2010. The proportion is much higher than that in Hehuachi Subdistrict (5.05%) because it was established as a base of personal services (32.89%) and retail trade industries (27.21%) in 2010.

In comparison, professionals and managers (38.5%) were dominant among all socioeconomic groups in Lianxin in 2010. During the 2000s, the absolute percentage change of professionals and managers (16%) was more than twice that of producer service workers, deriving from a cohort of new arrivals, with 6590 high-paid workers. Among these, managers consisted of merely 8.87%, compared with 29.63% professionals. Moreover, while managers in enterprises increased from 4.8 to 8.68%, those in both government sectors and public institutions decreased. Nevertheless, financial personnel accounted for only 4.2% of the employment in 2010, although this group experienced rapid growth. These situations recall the turbulence of labor retrenchment in state sectors and increasingly in private sectors caused by enterprise reform.

Although it has a higher share of manufacturing employment, Lianxin undoubtedly presented a lower share of low-paid production workers than did Hehuachi in 2010. This condition reinforces the argument that deindustrialization has emptied inner-city areas of workers but not necessarily of high-ranking persons in manufacturing enterprises. The percentage of commerce and service workers has increased in almost every inner-city subdistrict, except for the most gentrified area (Lianxin). In 2010, Lianxin accounted for 34.90% of all employment in low-skilled services (35.7% in 2000), while these workers consisted of 62.76% of the total in Hehuachi (43.28% in 2000).

Among all of these variables, the change in the housing tenure structure in the two subdistricts was the most pronounced, which explains the force of spatial commodification and housing consumption in gathering gentrifiers in certain places. The ratio of private homeowners in the most gentrified neighborhood changed from 5.84 to 49.91%, a far more rapid growth than that seen for professionals and managers. In absolute numbers, this result indicates an expansion of 10,320 households with homeownership in the open market (21,600 households in 2010). Public tenants in Lianxin decreased by 23.38% (2200 households), compared with a decrease of 16.96% in Hehuachi. Both areas contained only approximately 3% public tenants by 2010. Similar to the change in low-paid service workers, private tenants generally increased in the inner city following the housing reform. However, the number of new private tenants was much lower in Lianxin (1920 households) than it was in Hehuachi (6450 households). In 2010, Lianxin presented a high degree of homeownership for commercial apartments (49.91%) and a low share of private tenants (13.29%); the opposite situation was found in Hehuachi (16.04% and 32.42%, respectively). In line with the structural tenure change, rental payments by households in Lianxin in 2010 grew to more than eight times what they were in 2000, so this subdistrict shifted from being among the most affordable to being the most expensive inner-city area.

Finally, regarding the physical structures in place, by 2010, 82.87 and 67.79% of all households in Lianxin occupied housing constructed after 1990 and higher than 7 stories, respectively, compared with less than half in Hehuachi for both categories. In the most-gentrified subdistrict, the middle school score was the highest among all inner-city subdistricts in 2010. Residents in this subdistrict could enroll in two nationally ranked schools, two provincially ranked schools and one municipally designated school. In the least-gentrified subdistrict, however, residents could choose only between one provincially designated school and four municipally designated schools. These details thus confirm the proposition that, while the increase in inner-city producer service workers is a crucial indicator for the appearance of gentrification, currently, the gentrifier cohort in Chengdu remains characterized by a relatively mixed industrial structure but a common, high workplace ranking. This social condition emphasizes the homogeneity in economic capital but not necessarily an established cultural attribute among gentrifiers. Instead, the new urbanism and home-ownership, representing a form of cultural capital, could cultivate distinctions among the gentrifiers as they gentrify an area.

5.2 Consumer Class-Making

To understand the housing choice of inner-city gentrifiers, this study interviewed 20 residential gentrifiers in four newly built communities: Chengdunese Paradise, Waterfront, Times Riverside and Jintianfu. The first two communities could be treated as the product of commodity housing during the first round of urban redevelopment in the early half of the 2000s. Absorbing certain aesthetics of European and South Asian cities, the landscape of gentrification represented an early attempt of developers and local governments to materialize the dream of the good life for urban residents (Fig. 5.1). The last two communities were built during the second round of urban redevelopment at the end of the 2000s and are all located near Dongda Road at the southeastern corner of the inner city in areas previously dedicated to manufacturing (Fig. 5.2). The new residential spaces were advertised as introducing new forms of elite lifestyles associated with the increase in highly educated immigrants and young adults working in the new industries developed in the city.

The average housing price in the first two neighborhoods (which are nearly 20 years old) is approximately 25,000 Chinese yuan (approximately US$3870) per square meter, more than twice the average price in the main urban district of Chengdu. The residential gentrifiers moved into the first two communities over 10 years and are aged from 45 to 60, except one who is younger than 35. Most importantly, these gentrifiers emerging in the early 2000s showed common social ties with the public *danwei* system. They resided within either *danwei*-allocated housing or subsidized government housing before moving into this area. They are or once were cadres in the public sector or institutions. These families generally have very stable jobs. Except for the young gentrifier, they all possess multiple apartment units. Not a few of them purchased their current apartments without loans.[1] The other homes are located in

---

[1] When asked about the cost of the housing purchases and the debt situation, the gentrifier informants were usually reluctant to answer with exact numbers. Instead, they tended to compare the price range of apartments in their neighborhoods with adjacent neighborhoods to show the comparative advantages of the neighborhoods. However, some of them mentioned that they had no debt on their gentrified apartments, which were for residential use, but they may have borrowed to buy other investment properties.

Fig. 5.1 Gentrified landscape (Chengdunese Paradise and Waterfront) in the early 2000s. *Source* Photos taken by the author

either the inner city or the outskirts and either house family members or are rented out.

In contrast, the residential gentrifiers in the latter two neighborhoods were more often middle or senior managers in private enterprises than they were in state-owned ones and were more often professionals in schools, hospitals, financing institutions, etc. They are aged 35–60 and have resided in the communities for less than ten years. The family incomes range from 1 to 10 million. The families often raise more than one child, in contrast with the single-child families in the former two neighborhoods. The price of second hand housing in the nearby areas reached 30,000–40,000

Fig. 5.2 Gentrified landscape (Times Riverside and Jintianfu) in the late 2000s. *Source* Photos taken by the author

yuan (approximately US$4641–6188),[2] making this arguably the most expensive residential block in inner-city Chengdu.

5.2.1 Seeking "Authentic Urban Life"

Like many of the new homeowners in China, these inner-city gentrifiers gave primary settlement consideration to convenient facilities and beautiful and clean living environments, which generally indicate a high quality of community life [7]. The diverse and high-quality educational, medical, cultural, leisure and commercial facilities were attributed by the gentrifiers to the basic locational advantage of residing in the central city. In particular, the closeness to historic and cultural facilities and the symbolic meaning of inner-city places for Chengdu city identity constitute the most important factors that distinguish inner-city from suburban living. These

---

[2] In March 2021, the Chengdu Municipal Government issued the *Notice on Further Promoting the Steady and Healthy Development of the Real Estate Market*, formally intervening in the housing prices of the city. The notice strictly controls the highest house price in the main urban area of Chengdu below 40,000 yuan per square meter. The guidance price of second-hand housing in Jintianfu in 2021 is 30,466 yuan per square meter, lower than the actual market price of nearly 40,000 yuan per square meter before the introduction of the guidance price.

spatial resources, substantially and symbolically, not only offer convenience but also constitute what inner-city gentrifiers call an "authentic urban lifestyle."[3]

Historic places and buildings per se in China do not usually constitute an attraction for residential gentrifiers, but they offer opportunities for creative products and consumption experiences after renovation. Thus, as the middle class has presented a primary consumption inclination toward the new, or so-called modernist urbanism rather than historical features in housing design, living in the historic city enables residents to gain a sense of accessing Chengdu culture or the city identity. For instance, the teahouse, as a place where the Chengdu people meet daily and chat about politics, is an important component of the authentic Chengdu lifestyle [16]. The teahouse culture can be revived in historical areas in the inner city but in stylish ways rather than following traditional forms. However, the culture is largely disappearing in the newly established district in the south outside the Third Ring Road and being replaced by coffee shops. During the 2000s, residents usually made their first housing choice adjacent to the city core, but declining living conditions obliged them to ultimately select the high-end real estate initially developed in outlying areas of the inner city, such as the waterfront area and the near the Second Ring Road.

> We are not looking for a luxurious environment. We still like things with a little history, like I collect antiques. High-tech zones and eastern new areas are not like Chengdu. The paths beside the Qingshui River, Huanhuaxi Park, Du Fu Cottage and Qingyang Palace are the real Chengdu. It is the life of Chengdu people to invite a few old friends to drink tea from time to time. (G103, residential gentrifier)

5.2.2 Seeking Social Recognition

Notably, the informants were concerned with the social composition of the community, appreciating the homogeneity of the community members. This fact is underlined by the influence of social networking on gentrifiers' housing purchases. For instance, for informants from Chengdu Paradise, the gentrifier social networks were formed in either central city workplaces or intensively within the *danwei* system. Some of the gentrifiers thus selected communities where former colleagues lived. The other housing buyers could seek information from real estate brokers. Regardless of the channel, the gentrifiers were concerned with their neighbors, who are not only economic affluence but also highly educated and socioculturally minded.[4]

> The neighborhood can also be regarded as high end. The smallest apartment is 185 square meters, and the largest is more than 400 square meters. This size determines that the basic economic strength of the people who buy houses in the community is very good. Our underground parking lot is full of luxury cars, and many people wear clothes that look like luxury goods. Our property management fee is 5.8 yuan, which is very high in Chengdu. Property managers provide much better management of the community. (G113, residential gentrifier)

---

[3] Interviews with G93, 98, 103, and 102.

[4] Interviews with G93, 94, 100, 101, 103, and 105.

Housing selection based on social networking reflected the gentrifiers' internal need to establish place-based social recognition. However, for some of them, housing purchases were an approach to seize the ladder of upward mobility through housing consumption. This phenomenon was exemplified by two informants who had recently moved into their communities. The first community, Upper Town Chengdunese Paradise, was completed in 2009 and was the third phase of a large project that has been under construction since 2000 by the Dalian Wanda group. Mr. Chen (G98) is 52 years old and works as a manager in the public health sector in Chengdu. Chen moved from a small city due to a job transfer and then purchased a second-hand house in 2015 in Upper Town Chengdunese Paradise. The community was attractive to Chen due to its favorable social environment, with lifestyles compatible with that Chen had. In addition, his approval was based on public recognition of the district, which is widely acknowledged to be traditionally inhabited by people with high prestige and educational attainment. Living in the community facilitates social networking and career development for Chen; it is also where he gains a sense of territorial recognition and its privileges.

However, in attaining social status, Chen had to live in a second hand housing unit rather than a new one. Moreover, Chen paid a housing price in the community one-fifth higher than the average housing price nearby (1.7 million Chinese yuan for 104 m^2), which exceeded his budget. Chiefly, the high price was because the neighborhood contains a kindergarten and primary school with high educational quality available exclusively to community members. Chen was informed that residents outside the neighborhood would pay more than 100,000 Chinese yuan to have their child educated in the primary school. Since the late 2000s, educational facilities have become increasingly an important factor for stimulating housing purchases in inner cities. The educational policy in Chengdu has linked the community affiliation of residents (identified by one's household registration location) with educational opportunities to specific primary schools. The policy has encouraged residents to purchase inner-city apartments not necessarily to live in but simply to win a local *hukou*. This phenomenon has caused unusual price elevations for housing near well-known schools. In addition, located in the inner city, the gentrified area has been densely developed and is notoriously traffic congested, although within the community, high-quality green spaces, services and housing design are provided.

Researcher: Could you please tell me the reason you chose this community?

G98: We know many people who live in this district. In particular, this community is conventionally said to be high-end real estate in the inner city. The manager of XX Group (a large state-owned enterprise in Chengdu) also selected this community, even though he or she has lived in the United States for many years. When he or she returned, he or she purchased a secondhand house here. Near this community, on the opposite side of the road, another luxury community, which is invisible from the sidewalk due to a dark wall of trees, is sold only to people who have a specific rank of entitlement. We are not even permitted to visit it. This community is also rumored to contain many managers from other public sectors. My wife also considered the environment and the schools in this community. We even approached the question of schooling for our grandson in the future.

Researcher: Does this mean the price is very high?

G98: Of course, it is much higher than other communities that do not include schools. This school in our community is international. Moreover, there is heavy traffic in this district.

Researcher: So how do you feel about the neighbors now, after living in here for two years?

G98: It is still very good. I feel that the people in the community are mostly respectful, polite and elegant. Some of them know each other and can help each other become friends. Many of them have worked in public institutions, and the cultural level is good. We feel that most of the children are well educated, polite, and warm—not rude. (G98, residential gentrifier)

Another family living in Jintianfu consists of a middle-level male company manager and a female university professor. Both of them are 35–40 years old. They bought the smallest apartment available in this community, that is, 185 square meters. In the eyes of their colleagues, their living quality is far higher than the average. The couple reported that the home purchase consumed all their savings. At the same time, they borrowed more than 3 million yuan from their parents. More than three years have passed since they bought the house, and they are still trying to repay their parents. The purchase was expected to allow the entire family's quality of life and future financial and social status to move to a higher level. The male owner explained their initial thoughts about buying the apartment:

G110: We cannot afford to buy this house with our current economic strength, but we both thought about it and bought it. At that time, we felt that with the increase in housing prices, we might not have the opportunity to make a qualitative change in our living status in the future. Indeed, we seem to have entered a community that did not belong to us. Whether it is property management fees, heating costs or the cost of surrounding schools, we are still in the process of adaptation. However, I am very glad to have made this decision.

Researcher: What do you think about the difference between your household and the others in the community?

G110: One of the most obvious differences is that we have only one child, while other families have at least two and some have four. Excess children are registered as Hong Kong or foreign nationals. Almost every family has at least one nanny. Most of the families are business owners, and their income is naturally much higher than ours. The old Toyota I drive is probably the worst car in the neighborhood. These entrepreneurs seem to have different ideas from us. They are much more indifferent to each other and do not easily open their hearts. Some do not even want to be named. Is this the world of the rich? (G110, residential gentrifier)

5.2.3 Cultivating New Forms of Consumption

Residential mobility is cultivating gentrifiers' new consumption behaviors. The real estate market in China is still developing. At the same time, public service facilities and retails are advancing in synchrony with commodity housing construction in China's large cities. As a result, these gentrifiers may adjust and establish personalized consumption patterns after they move into their new communities.

The prominent commonalities in the daily consumption of the earlier gentrifiers are children's education and social consumption. These gentrifiers are inclined to adopt traditional styles of household goods and daily necessities—such as tea, antiques, calligraphy and painting—and are less interested in stylish goods or international brands. Some of them still retain the habit of using public transport. The new emerging gentrifiers have obviously begun to pursue high-grade, internationalized and customized consumption, such as fashion clothing and dolls and international brand home furniture. They also spend heavily on their children's education, personal health and care. However, compared with the former gentrifier group, who valued the high quality and strong security of public educational institutions, the new gentrifiers have a clear trend of choosing private and international educational institutions. At the same time, most of their children have received preschool education, such as equestrian, fencing, minor languages, and models.

> What we hope for now is to live the life we want to live, educate our children well, and live a comfortable and quality life every day. So as you see, we renovated this finely decorated apartment, as we did not like the vintage style of the developer's decorations. We like the modern style. It cost us 3 million to redecorate. We cannot wait to go home every day. Each piece of furniture was chosen by us. This table was approximately 100,000 yuan, and this lamp was 80000 yuan, all of which were imported from Italy. We truly want a home we like. Our biggest consumption is still our child. Our daughter is four years old. The cost of her interest classes is more than 200,000 a year, and her kindergarten is more than 100,000 yuan. However, my husband's clothes are all from Uniqlo (a cheaper brand). (G109, residential gentrifier)

In particular, the gentrifiers on the edge of the group are trying to adapt themselves to the new group by adjusting their lifestyles and consumption patterns. This process of cultural construction is typical. For example, Mr. Chen from Chengdunese Paradise has often treated friends to restaurants near the community and buys organic vegetables. After moving into their new home, the young couple in Jintianfu began to charge fitness cards, swimming cards, beauty and healthcare cards; to buy high-quality furniture and use high-quality household appliances; and to choose kindergartens with an annual cost of approximately 100,000 yuan for their children. At the same time, they must also adapt to daily living expenses such as gas and electricity, which could be twice as high as those in their previous residence. Of course, they often make trade-offs because of the increase in daily consumption, such as investing more in their children than in themselves, cooking at home rather than eating out, and renting or buying housing assets in relatively poor locations.

In addition, some recently gentrified communities provide one-to-one butler services. Each building is equipped with a housekeeper to help the property owners manage various family trifles. High security management also enhances the closeness of these neighborhoods. The property service fee is thus more than triple that of ordinary commodity housing. In enjoying the customized services of the neighborhoods, these gentrifiers may gradually form a higher sense of superiority, based on which they begin to form a higher level and quality of life. Therefore, changing consumption patterns, according to the gentrifiers, is not a way of showing off one's status but constitutes learned social activities to belong to the social circle to which he or she he aspires.

> People who live in Jintianfu should have some financial resources. Of course, these financial resources may also be given by parents. When we moved in, we could truly feel the superiority

of money. Every household has its own elevator; express is delivered to the door by a housekeeper; garbage cans are on each floor; water, electricity and gas fees are paid directly by the housekeeper; and all simple household maintenance is handled free of charge by the property manager. These changes can hardly not stimulate us, allow us start to reconstruct our daily consumption, to form a quality life. (G110, residential gentrifier)

5.2.4 Place-Based Identification and Differentiation

Identity refers to one's definition of self. Rather than a natural given, individuals develop identity in relationship to social and physical environments in a social world [11]. Identity development implies the commitment of individuals to a set of "commonly shared ... beliefs, rules, values, and expectancies" or "the statuses and roles of a given social group" [11, p. 58]. The branding and marketing process of a place creates and circulates symbolic meanings of consumption goods, which convey values, social statuses, beliefs, etc. [5]. Consumers then construct identities from those symbolic materials and internalize those meanings to account for themselves [15]. These gentrifiers may not have had obvious collective perceptions when they first bought an apartment, but, as they have constructed a new consumer culture and lifestyle, they have begun to have a sense of attachment to the residential area. Thus, they can generally distinguish the socioeconomic characteristics and lifestyles of different groups based on housing areas or types.

For the gentrifiers who emerged during the early 2000s, the stratum to which they belong or aspire is more consistent with a conventional definition of middle class, which is primarily characterized by a higher social or political status and educational attainment. Thus, they do not need great wealth but a stable income and career. Regarding the large emerging number of economically middle-class households in Chengdu, gentrifiers treat them as speculators whose wealth could be ephemeral, so their social status is less solid. When asked how they differ from the suburban population, the inner-city gentrifiers often questioned the disinterest of suburbanites in social and political affairs. They distinguished themselves from the residents in luxury suburban housing as having relatively lower personal wealth but possibly higher cultural capital, which leads them to appreciate the symbolic meaning of living in the inner city.

> For me, the middle class is first defined by income, and then, it must be concerned with cultural quality, somehow related to where you live. In terms of cultural quality, the ones you mentioned, who live in the luxury suburban housing, should not be described as middle class. Cultural assets include education, personal improvement and cultural taste; these are all capital. The real middle class has a stable income, which is much more important than how much money I currently have. Purely wealthy men are not a stable social group; they could be rich overnight but also go broke in one day. Moreover, this distinction is definitely related to one's educational level and cultural capital. However, of course, the middle class has a certain level of economic capital. For instance, they should be able to afford housing of more than 1 million in my opinion. However, someone with sudden wealth may choose luxury suburban housing, pretending that they are the upper class. However, we choose to be among the middle to high class. (G103, residential gentrifier)

For the new gentrifier generation, income is increasingly one of the core criteria for self-identification. Therefore, the important impact of housing on the identity construction of gentrifiers lies in its substantive value, including the location as well as the housing. Despite their occupational differences, gentrifiers share similar life values. They are busy with their work and do not have much time to care about political life. However, they care about their family lives and careers, and they always devote themselves to the development of small families or enterprises. They can be regarded as the backbone of urban economic development. In their way of doing things, they tend to be more pragmatic than idealistic. To a certain extent, they are comparable to suburban residents in housing occupation, income and culture, but their preference for urbanism distinguishes them.

G108: This is a very homogeneous area. Most of the households in this area of Chengdu should have attained a higher income level. These are people who do practical things—engage in business, make money and enjoy the present life—but we are not materialists. All in all, a realist.

Author: The West Qingyang District, south of the high-tech zones, is a good residential area, why do you not buy a house there?

G108: Qingyang District is the traditional elite area, where leaders in local public sectors often live. However, facilities in Qingyang District are relatively traditional, and housing facilities are rare. The overall development of Dongda Road is newer; it is a new rich area with more commercial facilities and the potential to be more internationalized in the future. We considered the southern high-tech zone, but our families like the atmosphere of Chengdu. The new district lacks place identification. (G108, residential gentrifier)

Author: Do you think that living here has affected your understanding of your social status?

G110: To a certain degree, it has had an impact. Housing is first of all a source of wealth. The growth potential of property prices in this area is greater than that in many other residential areas, which means that your wealth will increase. Location and the quality of the house, of course, location first, determine its value. In addition, you have this value, which can bring a certain sense of superiority. Of course, because consumption increases after we live here, it also stimulates us to continue to make money to maintain our status. (G110, residential gentrifier)

The above social processes reveal a constant formation and reconstruction of gentrifiers, ranging from the middle class directly or indirectly associated with the traditional *danwei* system to emerging private entrepreneurs and professionals. The emergence of new gentrifiers is also a process of constructing collective cultures and identifications based on spatial consumption. The cultural construction of gentrifiers in the two periods is also distinguished: the former is more traditional in consumption type, and the latter is more internationalized; the former is more conservative in lifestyle, and the latter is more avant-garde; the former is concerned with public affairs, while the latter focuses on private life. As the gentrification mode changes, new spatial and representative types might be produced, and the consumer cultures of gentrifiers are expected to be further divided.

Notwithstanding possible changes, compared with the other segments of the middle-class consumers in the city, the gentrifiers tend to distinguish themselves according to their urbanist tendencies, cultural concerns in their daily lives, and territorial attachments. In comparison with other parts of the city, the inner city is at the cutting edge of place-making for new urbanism, which is subsumed in the process of cultural globalization of the city. The new urbanism creates spatial capital, materially and symbolically, that assures a sense of advantage among gentrifiers selecting inner-city lifestyles. All of these concepts—namely, cultural advancement, authenticity, cosmopolitanism and elitism—tend to portray inner-city urbanism as a cultural product ahead of its time that will whet middle-class consumers' appetites for the sociocultural capital embedded in the renewed spaces.

5.3   Emancipation or Reproduction?

The social change in gentrifiers shows the different sociopolitics of gentrification in China from classical gentrification in postindustrial society. In the context of the consumption promoted by the reformist state of China, a body of literature has deployed the notion of the development of consumer citizenship to describe the characteristics of social change, in particular, from a sociopolitical perspective. "Citizen" discourse" in the public and political spheres in China coexists with the dissolution of discourses of the "masses" (*qunzhong*, 群众) or the "people" (*Renmin*, 人民) in a socialist collective society [1, 8, 14]. The development of consumer citizenship, then, embodies the practices of, on the one hand, the institutionalization of various rights and obligations of consumer subjects and, on the other hand, the internalization of consumer behaviors and identities (see [4]). While the Western literature has often articulated an idea of citizenship that enhances civil and individual rights, the literature in China has largely pertained to the new form of state–society relations in post-reform China (see typically [3, 4, 6, 13]). For instance, by reviewing media reports and national laws and by interviewing homeowners in a gated community in Shanghai, Davis [4] emphasized the characteristics of consumer citizenship development as the delegated authority of the state to balance consumer power with a low level of consumer activism in the political realm. Ren [12] analyzed the relationship between the middle class and the Chinese state in what he called the neoliberal era in China. He deemed the encouragement of middle-class consumer behaviors as a way for the Chinese state to transfer its economic responsibility to society, specifically, to individuals. Doing so both empowers and disempowers middle-class consumers by embedding entrepreneurship in society and by integrating economic risk into the lives of buyers and speculators.

The social status of gentrifiers adds evidence to the development of Chinese consumer citizens depicted by these authors. The social politics implied within the concept of emerging consumer citizens in China is in contrast with the cultural politics of the new middle class in gentrification in a postindustrial society. At least at the very beginning, gentrification in a postindustrial society symbolized a tide of cultural innovation and social liberalization promoted by a social transformation rooted in economic restructuring. First, classical gentrification led by urban pioneers revealed a critical social practice among the new middle class, i.e., a vision of an emancipatory city supporting a tolerant form of urbanism and a diverse urban culture (see [2, 9, 10]).

From the consumption side, the eagerness of Chinese gentrifiers for new lifestyle types also constituted a basic cause of inner-city gentrification in China. Ostensibly, gentrifiers' participation in the inner-city consumer market was conducive to cultural innovation in the city. However, Chinese gentrifiers are emerging in an era of consumer revolution and urban modernization. The sociocultural conditions inherited from the pre-reform cities have made it convenient for the politico-economic elite to implant their cultural imaginary into the city. These consumers are motived to pursue prestigious lifestyles that contribute to the decay of cultural otherness. The consumptive practices of those cultural consumers are concomitant with the state strategy of city consumer development and with the speculative nature of the real estate market, which together tend to sustain economic accumulation by selling new lifestyles to people in the city.

However, high-income consumers have not fallen into passive rather than creative roles in consumer culture. The development of a consumer consciousness has and will continue to create new demands among middle-class consumers as they seek new distinctions in habitus. Accordingly, urban regeneration periodically unearths a new consumer market based on the potential of inner-city areas to support cultural innovation. Once the cultural strategy and consumption stimulation take effect, gentrification will surge.

5.4 Conclusion

Part two uncovers two aspects of institutional change that characterize the state-facilitated gentrifying process in Chengdu. First, the state's urban imaginary and branding of the old city explains the changing cultural codes behind the place-making of gentrification. In Chengdu, the gentrification landscape of the 2000s reflected government ideas in the global city image and identity. The cultural codes changed from imitating the modernity in early developed countries to embracing minimalist modernism and, recently, emerging new ideas integrating Chinese traditionalism with modernism. The change happens in line with a growing international status of the city, which has increased the cultural confidence of the local governments. As the urban construction goals have changed, this urban imagination may increasingly move toward a compromise path between globalism and regionalism.

Second, the rule-making of land reuse on the brownfield sites explains the economic logic of gentrification in Chinese cities, which fits into the local government strategies that encourage excess production of spaces. In part, this phenomenon occurs because land marketization in China contains conflated economic and social objectives and signifies not only the expropriation of a putative rent gap but also the fulfilment of state legitimacy in a time of market transition. The overproduction of properties renders gentrification to be conditioned by the animation of a consumer market.

The new places produced based on the changing cultural ideas and economic rules are then embraced by emerging gentrifiers in parallel with the construction of their new collective cultures and identities. This study reveals a process of self-identification and self-distinction among inner-city housing consumers from other middle-class segments, who may choose a residence relatively distant from the inner city. In particular, historical and cultural resources and the cultural symbolism of authenticity and globalism provide scarce and territorially bounded resources, material or spiritual, that enable gentrifiers to pursue a sense of advantage and esteem by living in the inner city. The gentrifiers thus distinguish themselves from other middle-class segments by already-established social credentials and cultural capital in urban society, in addition to moderate economic affluence.

The orientation of social change thus signifies the formation of an ideological alliance between the new middle class and the politico-economic elites in the market society. As such, the gentrification process serves the grand engineering purposes of state-led social change and economic growth in a transitional society. Consequently, state-facilitated gentrification can hardly be expected to lead to emancipatory activities for sociocultural diversity. Instead, it leads to the construction of homogenous and conservative identity politics in a consumer society.

References

1. Anagnost A (1997) National past-times: narrative, representation, and power. Duke University Press, London
2. Caulfield J (1994) City form and everyday life: Toronto's gentrification and critical social practice. University of Toronto Press, Toronto
3. Davis D (2000) The consumer revolution in urban China. University of California Press, Berkeley
4. Davis D (2006) Urban Chinese homeowners as citizen-consumers. In: Garon S, Maclachlan P (eds) The ambivalent consumer. Cornell University Press, Ithaca, pp 281–299
5. Elliott R, Wattanasuwan K (1998) Brands as symbolic resources for the construction of identity. Int J Advert 17(2):131–144
6. Hooper, B. (2005). The consumer citizen in contemporary China. Working papers in contemporary Asian studies 12. Centre for East and South-East Asian Studies, Lund University
7. Huang X, Yang Y (2017) Urban redevelopment, gentrification and gentrifiers in post-reform inland China: a case study of Chengdu, China. Chin Geogra Sci 27(1):151–164
8. Keane M (2001) Redefining Chinese citizenship. Econ Soc 30(1):1–17
9. Lees L (2000) A reappraisal of gentrification: towards a "geography of gentrification." Prog Hum Geogr 24(3):389–408
10. Ley D (1996) The new middle class and the remaking of the central city. Oxford University Press, Oxford
11. Proshansky HM, Fabian AK, Kaminoff R (1983) Place-identity: physical world socialization of the self. J Environ Psychol
12. Ren H (2010) The neoliberal state and risk society: the Chinese state and the middle class. Telos 2010(151):105–128
13. Ren H (2013) The middle class in neoliberal China: governing risk, life-building, and themed spaces. Routledge, New York
14. Solinger D (1999) Contesting citizenship in urban China. University of California Press, Berkeley

15. Thompson JB (1995) The media and modernity: a social theory of the media. Stanford University Press
16. Wang D (2008) The teahouse: small business, everyday culture, and public politics in Chengdu, 1900–1950. Stanford University Press, Stanford, CA

Part III
Redistribution and the Post-gentrification Societies

Chapter 6
Redistribution

By examining the process of residential relocation in Chengdu, Part Three draws attention to the lower-income groups concerned in the gentrification process. It strengthens the previous understanding of displacement by unravelling the divergent experiences of lower-income residents during and after state-facilitated neighborhood redevelopment. The study focuses on three social groups: low-income workers living in public or subsidized housing; homeowners living in commodity housing, historical dwellings or self-built housing in the old neighborhoods; and migrant tenants. This Chapter first elucidates the policies and decisions for relocating and compensating the original residents impacted by inner-city redevelopment and documents the compensation results for the various social groups. These policies and decisions indicate an institutional change in housing low-income residents toward homeownership promotion and cultural assimilation. Additionally, the policies and decisions demonstrate the forceful continuity of both the property rights and *hukou* systems, which completely exclude migrant tenants from participating in the relocation plan (see also [41]).

6.1 Social Composition in the Redeveloped Neighborhoods

Work-unit (*danwei*) compounds and historical and dilapidated residential quarters were the two main dwelling types in the inner-city neighborhoods of Chengdu before the redevelopment, and of these, the *danwei* compounds constituted the largest proportion (Fig. 6.1a and c). *Danwei* compounds include housing built by work units and municipal housing authorities [43, 44]. Inner-city historical areas include dwellings dating back to Imperial China. The government confiscated the majority of these buildings when the Chinese Communist Party assumed power over

---

Parts of the findings from this chapter have been published in Urban Studies (2019, 56(12), 2480–2498), DOI: 10.1177/0042098018790724.

Fig. 6.1 Types of buildings in old neighborhoods: **a** public rental housing in CJA (above left); **b** commercial apartments in CJA (above right); **c** historic buildings in WNA (below left); **d** Urban village in the city (below right). *Source* Photos taken by the author in 2015 and 2016 and provided by a resident in WNA (6.1c)

Chinese cities. The government then rented these buildings to residents as public housing. However, certain buildings remained in the possession of individuals [16]. Other property types include a limited proportion of commercial real estate housing built during the early stages of the housing reform and ad hoc shelters built (often without full legal permission) by citizens or migrants. This research does not focus on the redevelopment of urban villages (Fig. 6.1d). Urban villages refer to settlements converted from rural villages during the process of rapid industrialization and urban expansion [18]. Instead of the inner city, urban villages constitute a main landscape of the urban–rural fringes in China's metropolises today [6, 42].

Each housing type share varies in each neighborhood investigated here. However, in general, long-established subsidized homeowners and public housing tenants together with private tenants constitute the principal groups affected by gentrification. This study describes them as socialist workers. Others include private owners, formally and informally, and migrant tenants.

Of course, this book does not intend to attribute one static attitude to *all* members of a social group. In fact, it is impossible to do so. On the one hand, resistance to urban redevelopment in China has repeatedly been found to fail as an organized, clearly targeted movement [23, 40]. Resistance is mostly individual- and case-based rather than collective, and the reasons for resistance are highly contingent upon family property circumstances. In the case of Guangzhou, Lin [17] revealed financial gains for landlords in a village that were derived from both the value of the compensation and the increase in the rental incomes following the greatly inflated housing prices after land redevelopment. Shin and Li [24], in contrast, emphasized the suffering of landlords who had illegally built additions to their properties from a loss of rental incomes due to the decrease in housing size after the compensation and resettlement. On the other hand, the resident reactions are variable. Even for activists, compensation that leaves them better off may diminish the primary motivation for resistance. I thus suggest that instead of documenting the immediate material gains and losses of individual households, we probe the social, cultural and subjective changes that the lower-income residents have experienced as a result of state-facilitated gentrification.

6.2 Homeownership Promotion and Rent Regulation Before 1990

After the urban redevelopment, households subject to compensation and resettlement consisted of only a part of the original residents. Migrant tenants were completely excluded (see also [17, 41]). The social results of gentrification cannot be simplified as the original occupants being dispossessed of their housing but as the state-driven reconfiguration of social power through urban redevelopment. It involves a process of the redistribution of the land conveyance fees and property increments among current and potentially new land users, which was closely related to the policy change of the housing system for lower-income groups.

During the 1980s and the mid-1990s, housing reform in China was understood as housing finance and management reform rather than direct housing privatization and marketization [34, 38]. The Chinese government intended to extend the financing channels of housing construction and improve housing quality for urban citizens rather than simply to retreat from the state provision of housing [1]. For instance, in the 1980s, the central state cut the funding of the housing provision to work units; meanwhile, fiscal reform increased their profit retention [1, 35]. The 1982 policy launched the method of an even tripartite sharing of a household's housing consumption by the state, *danwei* and households, although the practice was not always stringently implemented [1]. In both 1985 and 1998, the central government proposed constructing economical and functional housing, which would serve as the main form of affordable housing provision for the majority population in the lower–middle stratum after transitioning from a *danwei* allocation to a market system. Since 1991, the State Council has aimed to raise rents and established 'Public Funds for Housing' to promote current housing purchases.

An important result of the housing financing reform was that *danweis* increasingly became the main body financing and housing employees during the early reform stage; meanwhile, private funds have increasingly invested in housing construction. According to Bian et al. [1], *danwei*-led housing financing and provision have been based on several methods. First, wealthy SOEs raised funds and constructed/purchased housing for their employees. The funding channels were supported directly or indirectly by superior state agencies. Second, nonprofit institutions continued to gain housing funds from the state. Moreover, they were sometimes the primary land developers in coordination with secondary housing construction developers. Third, the most advantageous governmental agencies gained housing support from the real estate companies directed by the agencies. Fourth, among the private enterprises, wealthy ones (mainly international joint ventures) bought commodity housing at market prices, while less wealthy ones might not have provided housing for their employees. Instead, they depended on housing allocated from a previous work unit or to their spouses who were working in a state *danwei*. Last, the 'policy-permitted' deficit companies (see Bian et al.: 244) were subsidized by state housing funds, while other poor *danwei* either relinquished their housing provisions or handed part of them over to local government in exchange for municipal public housing.

The changes resulted in what Leaf [14] called *danwei*-led gentrification. In general, development companies with government agency or state-owned enterprise shareholders were the main actors in urban redevelopment in the 1980s and 1990s. High-ranked or rich *danwei*s and international joint ventures constituted the demand side of the housing market and probably also the supply side referring to housing employees. Meanwhile, economically eligible employees were encouraged to purchase subsidized housing from *danwei* or private commodity housing. When municipal public housing was redeveloped, public tenants were permitted to purchase municipal housing. At the same time, large *danwei*s continued to provide rental housing for low-income workers in *danwei*s after the redevelopment. *Danwei*s that were financially ineligible could transfer the responsibility for housing low-income workers to the municipal government based on payment of a certain amount; low-income workers were thus housed in public rental housing.

In this context of systemic housing reform, national and major city governments promulgated the first batch of urban housing demolition policies in 1991. Both national and local policies protected low-income local tenants at this stage. According to the 1991 *Regulation on Housing Demolition,* the Chinese government has attempted to maintain the supply of rental housing. The Regulation specified that renters were allowed to gain only in-kind compensation. Moreover, the policy clearly avoided the forced eviction of private tenants by property owners and required the stabilization of the private rental price before and after property exchanges. Similar regulations can also be found in the policies of Beijing, Shanghai, and Guangdong and Sichuan provinces. However, the home users who are entitled to the resettlement right in these regulations refer to residents who have permanent residence and formal housing without considering the compensation of migrants and informal housing residents [2, 7, 20]. These stipulations have effectively reduced the housing-induced deprivation of public and private renters residing in inner-city formal rental housing.

> The decision to demolish private rental housing (including other types of nonowner occupancy) shall be based on in-kind compensation for homeowners. The original lease between the homeowner and tenant remains unchanged after property exchange. Homeowners must

not force tenant evictions. When disputes occur over the lease, the parties can apply for mediation from the Housing Department or bring the case to the People's Court. [25]

As national policy stipulates the basic principles of demolition and resettlement, some local policies have begun to reflect the connection between compensation and resettlement methods and the promotion of housing reform. For example, the national policy suggests property exchange, pricing compensation or a combination of the above two as the main compensation methods. Private property owners and, mostly, managing agents of public housing (i.e., *danwei*s and government housing departments) can claim compensation [25]. Beijing's 1991 policy further stipulated that the resettlement of housing users should be carried out in line with local housing reform or in participation with the housing reform organized by the affiliated *danwei*s. The original users of demolished houses were encouraged to purchase resettlement housing [2]. The Chengdu policy is relatively obscure: "Demolition of urban public housing or private housing … after the property exchange … the original users are allowed to purchase the house to be resettled" [7].

Thus, *danwei*-led gentrification from the 1980s to the mid-1990s resulted in housing and tenure-based social differentiation among *danwei* employees. To a certain degree, housing-related social differentiation revealed the inheritance of the social power configuration based on the socialist work-unit system. The rich and highly ranked employees became homeowners as urban redevelopment proceeded, while the poor and low-ranked employees continued to live under the socialist housing system. In this process, the reconstruction of social space according to the difference in locational value began. Some urban residential areas have given way to the construction of public facilities and infrastructures. Smaller-scale *danwei*s were also likely to withdraw from the city center and to resettle their employees in the suburbs. However, owing to price control and rental regulation, housing displacement of the lower-income urban population remained limited.

6.3 Rental Deregulation and Dispossession After the Mid 1990

Since 1991, the World Bank has intervened in the structural adjustment of urban housing in China to promote housing purchases and rental adjustments. Mainly, it has continually questioned rental control by the government, which reduced the aspiration to buy residents, thus impacting housing value. The decrease in housing value ultimately generated a burden on work units who have to fund housing construction [33, p. 88]. Other questions include the shortage of credit arrangements and excessive government control over the housing system. In 1992, an implementation plan enumerated reform strategies for housing prices, urban planning systems, housing maintenance and mortgage systems [36]. In 1994, four cities were selected by the World Bank to facilitate the reform of the Enterprise Housing and Social Security System: Chengdu, Beijing, Ningbo and Yantai. Loans in the amount of $80 million were issued to Chengdu Municipality, compared with the $168.5 million issued to Beijing Municipality [37]. The World Bank suggested that the former methods of housing reform have failed to reach its goal, which was based on "gradual increases in rents while continuing to depend on enterprises, forced savings, and discounted sales of housing to bridge the gap between the rent and cost" (p. 2). The loan was released to encourage experiments with new reform methods in the selected cities to accelerate the full commercialization of the housing market, the complete restructuring of state-owned enterprises and the transition of the housing price system to a market base.

Driven by the transnational force, the State Council initiated the most comprehensive reform to date in 1994 and 1998. Reform in this period terminated in-kind housing allocation and replaced it with the establishment of housing subsidies and financialization [15, 27]. Employees' housing funds were derived from wages, public housing funds and mortgage loans. Meanwhile, the central government conceived of the establishment of a new housing system with a multilayered provision [26].

In the context of housing commodification, the urban redevelopment regime deregulated the rental market and robustly displaced lower-income residents from the inner city. First, an amendment to *the Urban Housing Demolition and Compensation Regulation* in 2001 no longer required the retention of urban rental housing but left the decision of retention or dissolution to the owners and lessees. However, at the same time, the stipulations required the property owners (i.e., *danweis* or government sector of public housing) to resettle the lessees (*danwei* residents) on the premise of terminating the lease relationship. Otherwise, the property owners could choose only to exchange property rights and maintain the lease relationship. The policy shows the central government's intention to advance housing commodification while urging *danweis* or housing sectors in localities to undertake part of the housing responsibilities at this time. In ambiguous language, the Regulation states the following:

> To dismantle rental housing, when property owners and tenants cancel the leasehold relation or property owners make housing plans for the lessee, only the property owners are the recipients of compensation. When the leasehold cannot be cancelled, in-kind compensation should be provided to homeowners, and the original leasehold relation should be maintained. ([28]; see also the policies in [3, 8, 21])

This Regulation, in effect, leaves decisions on maintaining leasehold relations to the property owners (see also [3]) but still attempts to protect tenant housing opportunities. The decision on compensation methods thereafter varied by city and project. Socialist public tenants who were formally eligible to claim compensation were commonly encouraged to purchase (see [39], [11] on Shanghai, [13] on Beijing). For private tenants, who are mostly migrants working in low-end industries in the central city, increasingly no formal policies have tried to protect their rental status. The policies that deem migrant vendors and tenants "illegal" and "informal" could cause local actors to be ignorant of the housing rights of the self-built homeowners and private tenants.

We do not need to take care of the floating population because they are illegal homeowners or non-property owners. Even citizens who built a squatter house by themselves still cannot be compensated as homeowners. We shall follow the rule and consistently maintain policy. Otherwise, property owners will think it is unfair. (O2, Official interview, 2014)

In both the 1991 and 2001 policies, subsidized owners (owners who purchased originally publicly owned housing) and private owners were offered two forms of compensation. They could claim property exchange (*canquan diaohuan*, 产权*diaohuan*) or price/cash compensation (*zuojia buchang*, 作价补偿 or *huobi buchang*, 货币补偿) on their own rights. The cash compensation was equivalent to the market value of the old property, while for the property exchange, homeowners were promised accommodations of an *equal size* after rehousing [2]. They could pay or receive the price difference between the original property and rehousing based on either the market price or the construction costs of rehousing. For the subsidized owners, their entitlement to housing welfare ended with compensation and resettlement [2]. Based on this policy, resettlement locations did not alter the unit size but could alter the price difference. Moreover, the location affected the rehousing costs of developers, which encouraged developers to choose peripheral districts for residential relocation. For homeowners of self-built houses, once their dwellings were designated "illegal constructions" or "temporary buildings beyond the legal deadline of existence" [28], they were permitted compensation only for construction costs rather than property rights.

These changes reflected a shift in the government's urban renewal purpose, from improving the living conditions of indigenous people to completely establishing the housing market mechanism. As a result, during the 2000s, urban redevelopment in Chinese cities transformed from a program that attempted to transition and maintain the ratio of affordable housing in the city before the mid-1990s to a central site shrinking affordable rental housing and resulting in displacement. In 2000, nearly half of all households in the main urban districts of Chengdu were either public tenants or owners subsidized by *danwei* and municipal housing authorities. Among the subsidized tenants and owners, over 75% of all households were located in the inner city. Throughout merely one decade (i.e., the 2000s), private homeowners of the newly built commercial apartments (not including subsidized owners of *danwei* housing) in the inner city increased from 6 to 30%. Accordingly, public tenants decreased from 23 to 2.5% in the inner city. In Shanghai, the number of original public tenants in the inner city decreased from 70 to 1%, while it decreased from 53 to 1% in inner-city Beijing (Population censuses, 2000, 2010). After 1998, national policy formally terminated the purchase of public housing by individual tenants. Thus, the large-scale decrease in publicly housed tenants has been the result of either gentrification or the spontaneous mobility of the previous tenants from the inner city.

6.4   Consumerism and Pro-homeownership Reform Since 2011

Indeed, the central government made attempts to establish a new affordable housing system. In 2007, national funds were issued for low-cost rental housing and price-controlled housing construction [29]. In 2010, the crucial New Ten Articles pioneered housing provisions dominated by social security dwellings and affordable commodity units [30]. In the context of the housing structural adjustment, urban redevelopment was promoted by the Chinese government as a strategy that simultaneously lead to housing improvement for the urban poor and market incentives. In the Executive Meeting of the State Council on 26 June 2013, Premier Li Keqiang introduced the urban redevelopment of dilapidated neighborhoods (*penghuqu gaizao*) as a focal point for the contemporary strategies of social and economic development in urban China. The urban redevelopment of dilapidated neighborhoods was explained as, on the one hand, a project for improving people's quality of life that would lead disadvantaged people into new housing and, on the other, economic growth through the promotion of domestic demand and consumption. The Chinese government has returned to its role as a social protector while incorporating this objective with a market incentive by promoting housing consumption by low-income groups.

In the meantime, a number of self-immolation incidents protesting forcible demolition and removal in the late 2000s called the attention of the national government to social stability in urban redevelopment [17]. In 2011, the State Council conducted an intensive review of the demolition regulations to clarify the power and responsibility of different parties and the basic principles and standards in the process of property expropriation. The main change of the 2011 *Regulation on the Conveyance of Buildings and State-owned Land and Compensation* (No. 590) lay in shifting the demolition agents from profit-oriented developers/demolition companies to government sectors or nonprofit institutions mandated by the government. Moreover, the notion of property *demolition* and *removal (caiqian)* in former policy discourses was terminated and replaced by property *conveyance (zhengshou)* and *removal (banqian)*. While property demolition was previously defined as acquisition and compensation conducted, sometimes violently, by demolition agents in the free market, conveyance means an *administrative* act of recovering state-owned land by state actors while legally transferring or cancelling the previous property registrations. Accordingly, the regulation abrogates the local government right to exert forcible tools of demolition and removal while assigning the task to the municipal legal rather than political system. Regulation No. 590 thus restored the principal role of the state in conducting housing conveyance and compensation. Therefore, urban redevelopment transitioned from a period featuring developer-led demolition and redevelopment to state-led conveyance and market-led redevelopment.

In terms of compensation methods, the No. 590 Regulation, for the first time, advocated nearby and in situ resettlement of expropriated households. However, this version of national policy did not provide specific compensation and resettlement measures for various real estate types, instead relegating those decisions to the subnational levels. The city level generally had two situation types. First, local policies began to explicitly carry out the public–private housing conversion by implementing urban redevelopment. Second, the local government continued to decentralize tenure conversion decision-making to stakeholders. Different from the 2001 policy, once an agreement was reached on tenure termination, an original tenant bought public housing at a discount and was then resettled by the developers instead of the original property owners (e.g., *danwei*).

For example, the 2011 *Regulations of Housing Expropriation and Compensation in Beijing* included a public housing expropriation and compensation. This section of the Regulation clearly notes that the lease relationship between the public housing owner and the lessee is automatically terminated from the issuance date of the housing expropriation decision [4]. However, public housing owners and lessees determined the treatment of the lease relationship in Shanghai and Chengdu [5, 22]. In these cases, because public housing assets have limited value and were bundled with many follow-up responsibilities for the lessees, most owners chose to terminate the lease relationship. The purchase price of the original public housing, which was far lower than the market price, also stimulated lessees to abandon the leasehold.

Decision-making by officials participated in gentrification projects in Chengdu shows vividly a path-dependent process of institutional change for housing the low-income *danwei* residents. In 2012, Municipal Secretary Huang Xingchu introduced the North Chengdu Redevelopment Program as a strategy for establishing the city as the growth pole of Western China. Before the central government most recently advocated for monetary compensation, the compensation policies of the North Redevelopment projects followed the 2011 Regulation, under which private owners could select either monetary compensation or property exchange.[1] Furthermore, compensation for public tenants was accompanied by public–private tenure conversion. Public tenants were obliged to pay a tenure conversion fee to the property's previous owners but at a heavily discounted price. Practically, however, the payment is deducted directly from the compensation fees for public tenants once they have been converted to private owners. However, private tenants, who are almost always unentitled rural–urban migrants (accounting for nearly two-thirds of the original residents in the study neighborhoods) are excluded from any type of compensation.

A direct reason for public–private tenure conversion is to win residents over to demolition and removal and to maintain social stability. However, essentially, the decision is a result of the retreat of both socialist employers and the local government from the responsibility for the housing welfare of work-unit employees. In Chengdu, according to the Municipal Housing Department, residents are allowed to conduct unified conversions from public tenancy to private ownership *only* through public projects, such as urban renewal work (O22).

---

[1] The new properties were located in the so-called resettlement communities. Some communities were constructed at the time of redevelopment, and the property tenures were not clearly defined, while other sources of resettlement housing were from established, price-controlled communities that were constructed before the redevelopment projects began. Some were located in the inner city or inner suburbs; others, at the distant outskirts.

In terms of work-unit housing, a change in tenancy follows the management regulations of the work unit and the administrative systems with which it is affiliated. Negotiations between the Housing Management Department and socialist employers, however, are not easy. On the one hand, *danweis* located in an urban district are often subject to an administrative group at a higher level than the local government.[2] The Housing Management Department needs to not only win support from the work unit at the locality but also to rely on the superior sectors of the administrative system that have decision-making power over the disposal of *danwei* assets. The overlapping jurisdictions of land and housing properties in the old neighborhoods is very likely to slow the negotiation progress (O2, O22, and O23).

The administrative sectors, however, are prudent in their decisions, as accepting public–private conversion means a drain of valuable assets. Paradoxically, old properties have long been a local danwei burden due to the annual expenditures on housing maintenance fees and housing-related welfare for residents, who were mostly retired or laid off. Maintaining the collective nature of the properties means having a constant postresettlement investment, whereas housing privatization is an effective way of quelling resident resistance and thoroughly relieving socialist employers from those responsibilities, albeit at the cost of losing collective property ownership.

> This is the dilemma of the old neighborhoods. They are of little value and yet not bad enough to be thrown away (*sizhiwuwei, qizhikexi*). However, when urban redevelopment occurs, the only choice you have is to cede property ownership. For if you did not, you would continue to have costs. It is impossible for *danweis* to provide the money; actually, they cannot even afford it. If you provided compensation for *danweis*, the residents would simply reject removal. Then, nobody could gain any profit. Thus, it is the top priority to ensure that the project successfully proceeds and then to maintain social stability. All of these objectives ultimately mean you have to offer profits to people (*rangli yumin*). (O35, manager of the *danwei* who owned the old properties)

The local government insists on the basic principle that it can compensate for property acquisition only rather than helping employers address their social responsibilities (O22). Conceivably, local officials and socialist employers deem housing demolition and compensation as a latent trigger that could incite a resident uprising. Ultimately, housing privatization, according to one official, becomes "the best way to both satisfy the housing demands of residents and terminate historical problems" (O22).

> We want to tell the work units that the properties they leave are generally negative assets. They are entangled in many historical problems. How can we do this (provide compensation)? The first way is to conduct public–private conversions, and then the historical problems will disappear altogether; another way, which is favored by the work units, is to let them first obtain compensation and then to help them tackle all of those problems. This is impossible. We can only compensate for the property, not answer questions. This is a difficult point. Ideally, they can gain compensation, as they are property owners. However, if the work unit gained the compensation, you left the burden on them. They have to cope with the burden. Thus, you divide their big cake into small pieces. (O22, official from the Urban–Rural Housing Department of Jinniu District)

---

[2] See Hsing [12, 35] for an explanation of the administrative system in China.

Another, less-noted, policy change is that the 2011 regulations at both the national and subnational levels defined compensation to previous property owners as being for the (market) value of the acquired properties rather than their size. The assessment of the compensation amount is thus based on a basic rule of the "equivalent exchange of properties or cash payment" (*dengjia zhihuan*, 等价置换). This fact means that "the amount of the evaluated market price of in-kind or cash compensation shall be equal to an evaluated market price of the old property" (O22). Based on this policy, the value of the in-kind cash compensation for each household is determined by the value only of the original property rather than by the rehousing location. However, the unit size of a rehousing is altered along with the resettlement site. Moreover, an appraisal of the value of old and compensated properties is conditional on the basic *marketable* prices of land in the old and relocated sites as estimated by the Municipal Land and Resource Department.[3] Thus, given the values of their previous homes, residents could barely afford new, onsite market housing of equivalent size unless a policy arrangement applied for onsite resettlement. Most of the residents who received monetary compensation could choose to buy new housing only in relatively poor locations.

In 2015, a new central policy encouraged monetary compensation (*huobihua anzhi*) for existing households in dilapidated areas (State Council [32], No. 37). This policy replaced the proposal for onsite relocation in the 2011 regulation [31]. The Ministry of Housing and Urban–Rural Development (MOHURD) and the China Development Bank (CDB) recently urged local governments to establish agencies and standards to assist in the transactions between previous residents in redeveloped neighborhoods and developers of unsold housing stock [19]. However, residents eligible for monetary compensation remain limited to homeowners or public housing tenants, not private tenants.

After Premier Li Keqiang advocated for neighborhood redevelopment, the Chengdu Municipal government, under the jurisdiction of the Housing Department in 2013, established an Office of Dilapidated Housing Redevelopment. After monetary compensation was found to be encouraging, former stipulations requiring the onsite resettlement of original residents were removed from policies (e.g., [9]) related to inner-city urban renewal. In addition, the Chengdu Municipal Government established a network for gathering information on the housing stock in the city and for assisting residents in purchasing housing post-gentrification with monetary compensation. Private developers with unsold housing were encouraged to participate in the program. Residents were allowed to use monetary compensation only to purchase housing or retail businesses. As a result, the proportion receiving monetary compensation increased from 53% of the eligible households in 2012 to over 70% in 2014 in Chengdu [10].

After decades of practice, the relocation and compensation methods of urban redevelopment have been increasingly institutionalized at both the national and municipal

---

[3] The price is calculated and standardized according to various conditions such as location and land-use category. It does not necessarily equal realistic market prices, but it does reflect differential ground rents.

levels. These policies show changes from the protection of public rental housing to deregulation to the promotion of housing privatization and consumption (Table 6.1). Housing reform along with urban redevelopment considerably accelerated the marketization of the housing provision for the local poor. Moreover, it resulted in a specific urban and social change in gentrification in Chinese cities: the further differentiation of housing provision for low-income residents and the attendant social restratification of low-income groups. On the one hand, public tenants and subsidized homeowners were thrown in sharp contrast onto the housing market, starting a new round of cultural transformation. On the other hand, migrant tenants are always excluded throughout the different processes of urban redevelopment. They suffered even more serious deprivations of housing opportunities as a result of the decrease in affordable rental housing after the old neighborhoods disappeared. Finally, private homeowners were obliged to accept the governments' arrangements for their housing, although they might have received the highest value of material compensation among the

Table 6.1 Policy evolvement of compensation and resettlement

	Compensation and resettlement methods		
	1991 policies	2001 policies	2011 policies
Public tenants	• Reservation of the lease and resettlement	• Decision-making decentralization: the encouragement of housing reform and resettlement by *danweis*	• Decision-making decentralization: the encouragement of public–private tenure conversion and resettlement by developers
Local private tenants	• Property exchange and the reservation of the lease	• Decision-making decentralization to private property owners	• Without compensation or resettlement
Migrant tenants	• Without compensation or resettlement	• Without compensation or resettlement	• Without compensation or resettlement
Homeowners of commodity housing	• Property exchange of equal size or cash compensation based on market price	• Property exchange of equal size or cash compensation based on market price	• Property exchange or cash compensation for equivalent market price
Subsidized homeowners	• Property exchange of equal size or cash compensation based on market price; cancel the entitlement to housing welfare	• Property exchange of equal size or cash compensation based on market price; cancel the entitlement to housing welfare	• Property exchange or cash compensation for equivalent market price; cancel the entitlement to housing welfare
Informal self-built homeowners	• Cash compensation for construction fees without resettlement	• Cash compensation for construction fees without resettlement	• Cash compensation for construction fees without resettlement

three groups. The next three chapters unravel the intricacies of the three groups' experiences.

References

1. Bian Y, Logan J, Liu H, Pan Y, Guan Y (1997) Work units and housing reform in two Chinese cities. In: Lu X, Perry EJ (eds) Danwei: the changing Chinese workplace in historical and comparative perspective. M E Sharpe, Armonk, NY, pp 223–250
2. Beijing Municipal Government (1991, Oct) Beijingshi chengshi fangwu chaiqianguanli tiaoli shishi xize (Detailed rules for the implementation of the regulations of Beijing municipality on the administration of urban housing demolition), No. 26. Retrieved 17 Dec 2021, from: http://www.beijing.gov.cn/zhengce/zhengcefagui/201905/t20190522_56276.html
3. Beijing Municipal Government (2001, Nov) Beijingshi chengshi fangwu chaiqian guanli banfa (Regulations of Beijing municipality on the administration of urban housing demolition), No. 87. Retrieved 17 Dec 2021, from: http://www.beijing.gov.cn/zhengce/zhengcefagui/201905/t20190522_56509.html
4. Beijing Municipal Commission of Housing and Urban-Rural Development (2012, July) Guanyu guoyou tudi shang youguan fangwu zhengshou yu buchang zhong youguan shixiang de tongzhi (Notice on the relevant matters in the expropriation and compensation of houses on state-owned land), No. 19. Retrieved 17 Dec 2021, from: http://www.beijing.gov.cn/zhengce/zhengcefagui/201905/t20190522_56509.html
5. Bureau of Urban-Rural Housing of Chengdu (2012, Dec. *Guanyu guoyou tudishang fangwu zhengshou yu buchang gongzuo zhong jige juti wenti de tongzhi* (Notification on several issues about housing conveyance on state-owned land and compensation). No. 131. Retrieved 22 Feb 2013, from: http://gk.chengdu.gov.cn/govInfoPub/detail.action?id=72493&tn=6
6. Cheng Z (2012) The changing and different patterns of urban redevelopment in China: a study of three inner-city neighborhoods. Community Dev 43(4):430–450
7. Chengdu Municipal Government (1993, Jan) Chengdushi chengshi fangwu chaiqian guanli tiaoli shishixize (Detailed Rules for the implementation of the regulations of Chengdu municipality on the administration of urban housing demolition), No. 30. Retrieved 10 Dec 2021, from: https://www.chinacourt.org/law/detail/1993/01/id/70882.shtml
8. Chengdu Municipal Government (2001, Nov) *Chengdushi chengshi fangwu chaiqian guanli zhanxing banfa* (Measures of Chengdu municipality on the administration of urban housing demolition and relocation), No. 88. Retrieved 10 Dec 2021, from: http://gk.chengdu.gov.cn/govInfoPub/detail.action?id=4527&tn=6
9. Chengdu Municipal Government (2013, Dec) *Guanyu jinyibu tuijin zhongxin chengqu jiucheng gaizao guifan fangwu zhengshou yu buchang xingwei de tongzhi* (Notification on promoting urban renewal in the central city and regulating housing conveyance and compensation), No. 57. Retrieved 10 July 2015, from: http://www.chengdu.gov.cn/wenjian/detail.jsp?id=ewAnYyOFWYfcUHepUf0E
10. China Business News (2015, May 1) *Zhujianbu tui "penggai huobihua": Fangdican you feng zhengce tuoshi* (Housing Ministry advocates "monetary compensation in urban renewal of dilapidated areas": real estate market boosting by policies). Retrieved 29 Mar 2016, from: http://finance.sina.com.cn/china/20150501/094622089205.shtml
11. Day J (2013) Effects of involuntary residential relocation on household satisfaction in Shanghai, China. Urban Policy Res 31(1):93–117
12. Hsing Y (2010) The great urban transformation: politics of land and property in China. Oxford University Press, New York
13. Kou Y (2013) The impacts of urban renewal: the residents' experiences in Qianmen, Beijing, China (2013). Dissertations and Theses. Portland State University
14. Leaf M (1995) Inner city redevelopment in China. Cities 12(3):149–162

15. Lee J, Zhu Y (2006) Urban governance, neoliberalism and housing reform in China. Pac Rev 19(1):39–46
16. Liang SY (2014) Remaking China's great cities: space and culture in urban housing, renewal, and expansion. Routledge, New York
17. Lin GCS (2015) The redevelopment of China's construction land: practising land property rights in cities through renewals. China Q 224:865–887
18. Liu Y, He S, Wu F, Webster C (2010) Urban villages under China's rapid urbanization: unregulated assets and transitional neighborhoods. Habitat Int 34(2):135–144
19. Ministry of Housing and Urban-Rural Development and China Development Bank (2014, Oct) *Guanyu jinyibu jiaqiang tongchou xietiao yonghao penghuqu gaizao daikuan zijin de tongzhi* (Notification on efficiently using bank loan funds for the renewal of dilapidated areas in cities), No. 155. Retrieved 20 Aug 2015, from: http://zfbz.cfjianshe.gov.cn/index.php?m=Show&a=index&cid=510&id=565
20. Shanghai Municipal Government (1991, June) *Shanghaishi chengshi fangwu chaiqianguanli tiaoli* (Regulations of Shanghai municipality on the administration of urban housing demolition). No. 4. Retrieved 19 Dec 2021, from: https://www.oh100.com/a/201610/411112.html
21. Shanghai Municipal Government (2001, June) Shanghaishi chengshi fangwu chaiqian guanli shishi xize (Detailed rules regulations of shanghai municipality on the administration of urban housing demolition). No. 111. Retrieved 19 Dec 2021, from: http://www.jingan.gov.cn/xxgk/016052/016052014/016052014006/20011101/d5119a29-2362-41a2-b6ba-5c31bda885ad.html
22. Shanghai Municipal Government (2012, Oct) Shanghaishi guoyoutudi shang fangwu zhengshou yu buchang shishi xize (Detailed rules regulations of Shanghai municipality on the expropriation and compensation of houses on state-owned land), No. 71. Retrieved 17 Dec 2021, from: https://www.66law.cn/laws/114616.aspx?winzoom=1
23. Shin HB (2013) The right to the city and critical reflections on China's property rights activism. Antipode 45(5):1167–1189
24. Shin HB, Li B (2013) Whose games? The costs of being "Olympic citizens" in Beijing. Environ Urbanisation 25(2):559–576
25. State Council of the People's Republic of China (1991, Mar) *Chengshi fangwu chaiqian guanli tiaoli* (Ordinance of the administration of urban housing demolition). No. 78. Retrieved 9 July 2015, from: http://law.npc.gov.cn/FLFG/flfgByID.action?txtid=2&flfgID=11993&showDetailType=QW
26. State Council of the People's Republic of China (1994, July) *Guanyu shenhua chengzhen zhufang zhidu gaige de jueding* (Decision on deepening the reform of urban housing system), No. 43. Retrieved 17 July 2015, from: http://www.gov.cn/zhuanti/2015-06/13/content_2878960.htm
27. State Council of the People's Republic of China (1998, July) *Guanyu Jingyibu shenhua chengzhen zhufang zhidu gaige jiakuai zhufang jianshe de tongzhi* (Notification on further deepening the reform of the urban housing system and accelerating housing construction), No. 23. Retrieved 17 July 2015, from: http://www.ggj.gov.cn/vfggw/qtfg/200806/t20080610_262964.htm
28. State Council of the People's Republic of China (2001, June) Chengshi fangwu chaiqian guanli tiaoli (Ordinance of the administration of urban housing demolition), No. 305. Retrieved 9 July 2015, from: http://www.gov.cn/gongbao/content/2001/content_60912.htm
29. State Council of the People's Republic of China (2007, Aug) *Guanyu jiejue chengshi di shouru jiating zhufang kunnan de ruogan yijian* (Suggestions on solving housing difficulties for urban low-income households), No. 24. Retrieved 16 July 2015, from: http://www.gov.cn/zwgk/2007-08/13/content_714481.htm
30. State Council of the People's Republic of China (2010, Apr) *Guanyu jianju ezhi bufen chengshi fangjia guokuai shangzhang de tongzhi* (Notification on decisively controlling the overly rapid increase of housing prices in some cities), No. 10. Retrieved 16 July 2015, from: http://www.gov.cn/zhengce/content/2010-04/17/content_4639.htm

31. State Council of the People's Republic of China (2011, Jan) *Guoyou tudi shang fangwu zheng-shou yu buchang tiaoli* (Ordinance of housing conveyance and compensation on state-owned land), No. 590. Retrieved 10 July 2015, from: http://www.gov.cn/zwgk/2011-01/21/content_1790111.htm
32. State Council of the People's Republic of China (2015) Opinions on the improvement of dilapidated housing renewal and facilities construction, No. 37. Retrieved 21 February 2016, from: http://www.gov.cn/zhengce/content/2015–06/30/content_9991.htm
33. Tolley GS (1991) Urban housing reform in china. The World Bank
34. Wang YP, Murie A (1996) The process of commercialisation of urban housing in China. Urban Stud 33(6):971–989
35. White G (1993) Riding the tiger: the politics of economic reform in post-Mao China. Stanford University Press, Stanford, CA
36. World Bank (1992) China: implementation options for urban housing reform. World Bank, Washington
37. World Bank (1994) China enterprise housing and social security reform project. World Bank, Washington
38. Wu F (1996) Changes in the structure of public housing provision in urban China. Urban Stud 33(9):1601–1627
39. Wu F (2004) Transplanting cityscapes: the use of imagined globalisation in housing commod-ification in Beijing. Area 36(3):227–234
40. Wu F (2004) Residential relocation under market-oriented redevelopment: The process and outcomes in urban China. Geoforum 35(4):453–470
41. Wu F (2016) State dominance in urban redevelopment: beyond gentrification in urban China. Urban Aff Rev 52(5):631–658
42. Wu F, Zhang F, Webster C (2013) Informality and the development and demolition of urban villages in the Chinese Peri-urban Area. Urban Stud 50(10):1919–1934
43. Zhang XQ (1997) Chinese housing policy 1949–1978: the development of a welfare system. Plan Perspect 12(4):433–455
44. Zhang Y, Fang K (2004) Is history repeating itself? From urban renewal in the United States to inner city redevelopment in China. J Plan Educ Res 23(3):286–298

Chapter 7
Cooperation

7.1 For a Better Life

The government policy of homeownership promotion has significantly affected the attitudes of public tenants and subsidized owners regarding gentrification projects. Among the affected residents, these were the only social groups that did not necessarily question the rationale of urban redevelopment and relocation. They were once low-paid socialist workers and were registered as local citizens; they had been housing welfare recipients since the pre-reform period. The Spartan housing system in socialist cities and the lack of affordable housing for low-income residents in the new housing market preconditioned these residents to housing impoverishment and caused them to place their expectations in urban redevelopment and relocation.

> Our demands are simple: the compensation unit can accommodate all the family members and ensure functional separation between generations. We wish to live in a partitioned apartment. We have used a common kitchen and washroom for our whole lifetime. It is inconvenient and dirty. (R11, resident relocated offsite)

Nevertheless, a precondition for housing impoverishment has been the Spartan housing system in socialist cities and the shortage of housing for low-income residents in the new housing market. Until the end of the 1990s, when central policies terminated housing allocations based on *danwei*, residents were allowed to request that socialist employers supplement their housing provisions. In most cases, low-income citizens periodically claimed an alternative subsidized or allocated unit but only when their families expanded or their children reached adolescence. During the 2000s, new housing construction was dominated by commodity rather than affordable housing, and all housing price types skyrocketed in urban China [3]. This situation further hindered housing improvement for low-income citizens. Since 2010, the central government has advocated a new low-income housing provision system in post-socialist cities. The fate of such advocacy remains to be seen, especially

This chapter has been rewritten and published in Urban Studies (2019, 56(12), 2480–2498), DOI: 10.1177/0042098018790724.

© Springer Nature Singapore Pte Ltd. 2022   153
Q. Yang, *Gentrification in Chinese Cities*, Urban Sustainability,
https://doi.org/10.1007/978-981-19-2286-2_7

concerning the insurmountable problems associated with central–local differences and fiscal obstacles.

> Researcher: Did you ever plan to buy a commodity house?
>
> R6: Salaried people like us can by no means afford commodity housing in our lifetimes, unless with government support.
>
> Researcher: Why didn't you apply for low-rent housing?
>
> R6: Some have indeed applied, but more for their children. Actually, you have to wait for several years, especially when you have no social relationship (*guanxi*). There are many conditions (for application) that nobody fully understands. Sometimes they ask you to provide a certification from your work unit, but I have not been subject to any work unit. I think the Residential Committee should provide certification for people like us. There are always various difficulties when poor people want to do something. Moreover, if it is still rental housing, who would like to move outside of the Third Ring Road? (R6, resident relocated offsite)

Ultimately, overwhelmed by the deficiencies of the new housing security system, the soaring housing prices and the fervor of urban redevelopment, low-income home-owners and public tenants have been forced to place their expectations in so-called demolition and relocation (*chaiqian*). Moreover, the special policies of homeown-ership promotion have offered them a welcome opportunity for state-subsidized homeownership: "We have to wait for a long time for demolition. If it were not for the compensation policy, we would no longer flee from this dilapidated place" (R5).

In addition to the housing demand, the path dependence of the state–*danwei* worker relationship, shaped in the pre-reform era, impacted the participation of this social group in urban redevelopment. During the planned economy, workers in the urban state sectors in China were trained in loyalty to the party state because party membership was integral to personal performance in *danweis*. Still, these workers were deeply dependent on state sectors because the labor force was extremely immobile and the work units had an exceedingly close community life [12, 13].

The ideological legacy of collectivism might sometimes subject individuals to certain types of collective interests. For instance, in the case of an overlapping house-hold head for two physically separate properties, a redevelopment norm regulates that the head can be compensated only according to the principle of a single household. The norms were instituted to control speculative behavior. A resident encountering this situation expected compensation for both independent properties so his child's family and his family could live separately. He then endured ideological struggles:

> R5: Actually, I support the redevelopment. However, the demolition office did not treat the two units independently. Frankly, I had a strong conflict in my mind.
>
> Researcher: What do you mean by a strong conflict in your mind?
>
> R5: I think it is my right to claim two compensation properties. However, they made a Caojia Alley policy in which properties under one head were integrated as single household, regardless of the area size.
>
> Researcher: Therefore, you think the policy is reasonable?
>
> R5: Yes. I definitely received a mass. Considering the situation, I could only give up contending for my own benefits. (R5, resident relocated onsite)

7.2   The Status Change of Socialist Workers

Socialist workers tend to treat urban redevelopment and compensation as a critical vehicle for change in their lives: "Please feel free to demolish it; you demolish once, my life gets better. The more you demolish buildings, the better my life becomes" (R17). Thus, even in the case of relocation offsite, from the very beginning of the policy's implementation, socialist workers were more likely to interact with officials in patterns of cooperation and negotiation rather than resistance. Moreover, in the later negotiation stages, this social group has played an important role in consensus-building over demolition and removal. Although dissidents remain, their arguments mostly pertain to the fairness of the compensation and property evaluations and the construction and site selection of the resettlement communities rather than the legitimacy of demolition and relocation.

7.2.1   The Advantages of Possession

With a soft attitude, socialist workers overtly sought in-kind and monetary gains by participating in redevelopment projects. State-subsidized homeownership after redevelopment not only ensures socialist workers a new and higher-quality home but also results in the anticipation of financial benefits realizable in the market. Before redevelopment, the old CJA neighborhood was the most densely occupied section of the city. Built by the No. 3 Construction Company of Sichuan Province in 1956, public rental housing in CJA is typically characterized by two-to three-story, redbrick blocks of a minimalist and standardized design. The structures, referred to as barrel buildings (*tongzilou*), resemble studio apartments but of far lesser quality. The basic housing unit is a single room of an average 16 m^2, which, in extreme cases, accommodates two generations of four to five members. Rooms are positioned along corridors, within which residents, who also share public toilets, must cook (Fig. 6.1a).

In the CJA redevelopment project, socialist workers were compensated for their public/subsidized housing with an average offer of 8000 Chinese yuan (approximately US$1231) per square meter of the old property. The government offered discounted prices of 4500 Chinese yuan per square meter (US$692) for offsite resettlement units and 7000 Chinese yuan (US$1077) for onsite resettlement units as a concession to residents selecting new properties (O2, official from the CJA Residents' Committee). When other subsidies are considered, a two-generation family previously living in a two-room unit of 58 m^2 were able to exchange it for a nearby home of 90 m^2 priced on the open market in 2013 at approximately one million Chinese yuan (US$153,846) by paying less than 20,000 Chinese yuan (US$3077) in a tenure conversion fee (R4, CJA resident relocated onsite). The offsite rebuilt apartment could be 1.5 times larger than the onsite option because of the difference in housing prices.

Aside from the investment potential, the informants also valued the high level of individual autonomy they had postresettlement. In personal narratives, they regarded Mao's China as frustrating their path to success and keeping them impoverished. Frequently mentioned topics were the plights of families as a result of the political suppression experienced during the Cultural Revolution; the unfairness of the danwei bureaucratic system regarding career changes and housing welfare; and barriers to personal development, such as schooling and employment, due to political favoritism. Faced with all of these daunting experiences, the informants anticipated transcending their pasts by embracing a new identity as self-governing, home-owning citizens.

Membership in a gated community signifies the termination of the state–*danwei*–worker relationship and a corresponding enhancement in individual freedoms in organizing one's personal work and social life. As argued by Pow [7], living in the new, gated communities means not only a retreat to privately developed spaces but also an enhancement of 'household autonomy and personal freedom,' in contrast to 'the hegemonic control of the Communist Party state' (p. 830).

> Now we have a sense of autonomy. This is (due to) what they called 'returning the power to the people' (*huanquan yumin*). If you have nothing, where is the power? Now you are the property owner. You gain discursive power. Now the Street Office, rather than *danwei*, keeps my archive. Nobody can watch it and change it informally. (R10, resident relocated offsite)

Therefore, homeownership also increases self-satisfaction and standing in a market society [4]. For instance, a new apartment may enable a young man in a relocated family to enter into marriage or a young couple to begin a family postresettlement. The older generation expects to leave a valuable legacy, which will help their children's families break the poverty cycle.

> Indeed, this will be my son's house. It will be his capital to develop his family after we all have gone. For this reason, I selected onsite relocation. The offsite option could have earned me one more apartment, where we could live separately, but the onsite apartment will leave them more wealth in the near future. (R4, resident relocated onsite)

Ultimately, perceiving these status advantages, nearly all of the onsite resettled residents and a portion of those resettled offsite indicated a sense of upward mobility in social status in the resettlement communities. Subjectively, the phrase 'propertied stratum' (*youchan jieceng*) was commonly invoked by the informants to draw a contrast with their former proletariat status.

Compared with the public tenants, the former owner-occupants of subsidized housing experienced relatively stronger prior recognition as homeowners. They tended to consider the compensation available to them as less satisfactory than did the tenants, as property exchange without tenure change results in less value added. However, having formerly been subject to the work-unit system with relatively low incomes, the owner-occupants still valued the larger sizes of their newly subsidized housing both functionally and as an enhanced asset.

The experiences of the public tenants and subsidized homeowners indicated that they recognize gentrification premised on an existing ideology of possession. This argument echoes [10] observation on the pivotal role of distinctive property and tenure systems at the local level in contextualizing gentrification. However, the argument takes on a slightly different character in a society like socialist China that lacks a completely transformed land and housing market, where "dispossession of [the original residents'] right to properties and to the city" is a precursor to gentrification [9, p. 484]. We suggest that, from the diverse perspectives of the residents, a sense of being dispossessed from the right to the inner city was more obvious among the private homeowners of both commercial apartments and self-built housing in urban villages when they faced expulsion from a redeveloped neighborhood than it is among the self-identified proletariat. When the compensation policies purposively enabled an opportunity to "own," the attitudes of the previous proletariat diverged from those of existing private homeowners. As the residents commonly acknowledged, "generally households with homeownership do not want the old housing to be demolished, while those without private housing hope for demolition and resettlement" (R7). Therefore, gentrification in China might not only presuppose the market regime dispossessing accumulation but also, as we shall see, societal reconstruction toward a fully commodified system in everyday subjectivities.

7.2.2 Lifestyle Change

Since the 1990s, property developers have introduced new dwelling types to mold an urban lifestyle distinct from the socialist era, notably gated high-rise or multistory dwellings with a central garden or luxury suburban communities [8]. Concomitantly, private consumption prevails; residents have moved into new apartments that distance them from the socialist-era communal society. These habitats comprise semantic fields with new social and cultural norms [1, 11]. Lifestyles in gated communities are characterized by diversified personal consumption, an expansion of private life, and the social exclusion of other housing classes [7, 8, 15]. The socialist workers' sense of social elevation may appear cosmic: 'The resettlement indicates our transformation, an upgrading and a residential transformation. Previously, we were residents; now we are proprietors. The change is from the underground to the heavens' (R29, resident relocated onsite) (Figs. 7.1 and 7.2).

The relocated residents insisted that a major novelty was the increase in personal consumption. Because their recreational facilities, sanitation, and security are provided by the private developers and property managers, their main personal investment is home decoration. An elaborately decorated apartment, which sharply contrasts with the single room in a *danwei* compound, is a popular means of displaying lifestyle change. In other cases, however, the relocation and compensation stimulated both unselfconscious and conspicuous consumption, for example and in particular, of personal vehicles among the younger generation.

Previously, beds lay one on top of another in the room. There was no concept of decoration. It was a public asset. Now, this is my own apartment. You see the kitchen is spacious; the color is light. See my daughter's bedroom. The purple curtain facing the central garden, the

Fig. 7.1 Onsite and offsite resettlement neighborhoods. *Source* Photographs taken by the author between 2014 and 2015

Fig. 7.2 Jinniu Wanda before and after redevelopment and onsite resettlement. *Source* Photographs taken by the author between 2014 and 2015

pink wallpaper and the white wardrobe and bedding are all colors that she likes. We can now finally realize her dream space. (R12, resident relocated offsite)

After the resettlement, daily interactions among the residents in public spaces were greatly reduced, with a corresponding increase in family-centered life. Meanwhile, unlike the old expanded social networks among neighbors and *danwei* coworkers, social life in the resettlement communities swiftly shrunk to close friends. The residents attributed this social transition to the privacy accompanying the separation of neighbors in high-rise apartments.

The private space in our former compound was too small. Most of the daytime we remained in the corridor, cooking and talking with each other. We participated in classes on Chinese calligraphy in the workers' cultural center. Now, however, our connection is increasingly weakened, even with former neighbors who relocated to the same community with us. There are some changes in feelings between us. We have only a few close friends left. We dance in the neighborhood garden each day after dinner (R16, resident relocated onsite) (Fig. 7.3).

Fig. 7.3 Life within the onsite (left) and offsite (right) resettlement communities for the CJA residents. *Notes* Open streets with retail stores were designed for the onsite resettlement community in an attempt to reproduce the street spaces of the old CJA. *Source* Photographs taken by the author between 2015 and 2018

7.2.3 New Social Organization and Exclusiveness

The resettlement communities have experienced a social reorganization, as these new proprietors cultivate exclusivity. Two years after the resettlement in the JW neighborhood, the property manager urged the socialist workers to establish a homeowners' association. Accordingly, the Residents' Committee, which previously governed the *danwei* habitat and undertook tasks assigned by the local government, retired from community management. Socialist workers have learned to behave like proprietors, assuming self-government and participation as well as acting for personal profit. The gated community has also permitted the relocated residents to separate themselves from other social groups with different habitus. Exclusivity is overtly manifested in the relationships between homeowners and tenants and between local residents and migrants. In the JW resettlement community, households that own a second property may lease it to migrant tenants to generate extra income, a practice known as group renting (*qunzu*). In addition, tenants, as petty businessmen, may use the apartments commercially for restaurants or storage. The group renters and their business activities deeply frustrate the homeowners, who are concerned with community security, residential quality, the social environment, and facility maintenance. In the homeowners' association, socialist workers have mobilized to establish measures to curb group renting to migrant tenants and unreasonable nonresidential apartment use.

> This [resettlement] community is near a large wholesale market and a railway station. Approximately 70% of the community used to consist of outsiders and renters. They used the apartments as storage or as self-employed business locations. The city residents barely get along with the rural people. They are not rational and are poorly educated, merely expressing personal interests or that of their small groups. They do not know how to participate in community management. (R15, resident relocated onsite)

Analyzing social stratification in a market economy, Weber [14] separated the concepts of social status from economic class. While class is determined by the economic resources a social group can muster in the market, status takes shape in the social sphere and is embodied in an assumed lifestyle. Attaining an elevated status invokes a specific honor or achievement and impacts social networks of interaction and exclusion. The possession of domestic property announces the social transformation of socialist workers, embodied in a status change and a sense of upward social mobility through the fact of possession, changing lifestyles, and the cultivation of social interaction patterns that include exclusiveness as members of newly built, gated communities.

7.3   The Uncertainty of Social Transformation

Weber [14] defined the possession or lack of property as 'two basic categories of all class situations' (p. 182). The good fortune of socialist workers in urban redevelopment situations comes from 'obtaining a very important thing that originally the workers could by no means afford' (R35, resident relocated offsite). However, home-ownership must generate a market return to ensure a qualitative change in economic class. Accordingly, a few JW informants converted their apartments to mixed use as both residences and self-employed businesses. Some CJA residents purposely selected offsite relocation to gain more than one property. The benefits generated by renting or selling the extra apartments were then invested in personal skills, such as learning to drive, thereby increasing job options. These initiatives appeared in families in which the young generation aggressively used the homeownership opportunity to expand personal mobility.

Nevertheless, the majority of older and underemployed families still live on pensions or a government allowance. Moreover, because of state welfare was retrenched in favor of housing privatization and commodified redevelopment, these new homeowners, especially those remaining in the inner city, have confronted increasing consumption pressure. Living in the JW community near an ostentatious shopping mall, the onsite resettled residents rarely purchase daily goods nearby, instead turning to less expensive services in old neighborhoods. Moreover, after losing their ability to rely on the government and previous employers, former welfare recipients must now bear their own housing-related expenditures, and they may be hard pressed financially even before the inevitable costs of home maintenance. After fruitless requests to property managers and developers for repairs, helpless residents may return to the government.

> Many people cannot even afford the property management fees. Some have resisted paying them. Some have paid them, but they said because they paid, the property management company should be responsible for all maintenance costs. However, housing maintenance is indeed troublesome. The construction company took only half a year to finish the construction of our community. The wall is rough; the window is slanting. I have repaired sewer leakages four times during the past half year. I cannot afford it if it needs to be constantly repaired. Do you know who I should approach? (R38, resident relocated onsite)

Thus, the social status change may not ultimately advance the life chances of all of the socialist workers. Indeed, the consumption pressures on the onsite relocated residents in the inner city may even lead to out-migration. Some of the residents have sold their resettled onsite apartments and purchased secondhand units relatively far from the inner city. Others might reuse the apartments as hotels or shops or subdivide the apartments into small rooms that they then rent separately. Some of the resettlement communities have evolved into places that absorb welfare-deficient urban citizens, underemployed farmers and new incoming migrant tenants. This situation raises another series of problems concerning community disorder and management challenges. Thus, the promotion of working-class homeownership may result in resettled sites, such as the old neighborhoods, being disinvested and unevenly developed. Will the disorder in the resettlement communities symbolize another round of poverty concentration? Two years after moving to the JW resettlement community, a voluntary mover reassessed the choice:

> Sometimes I feel it may have been better if we had stayed in our former home. They would repair the sewage system, toilets and kitchens for us. We would not feel so lonely, helpless; also, there would be no troublesome issues in the community. People who are still expecting wealth from demolition will one day know the situation once they are resettled in a new place. (R33, resident relocated onsite)

Finally, even though the compensation scheme has enabled formerly public tenants to become homeowners, at present, official policies do not clearly define the tenure type of resettlement housing. The definition has remained flexible so the state can rearrange the public or private natures of the housing and its attendant property rights. Thus, certain spaces in Chinese cities have quite obscure public or private attributes (e.g., resettlement communities for urban populations or land-lost farmers). This evidence confirms that the ambiguity of the property rights system allows leeway for the Chinese state to arbitrate between the market and society [2, 6]. Previously, Lin [5] found that the municipal government in urban redevelopment has endeavored to define "who gains what rather than who owns what." This study reveals that the decision-making on gains is subject to change according to variable definitions of rights attached to tenure types.

References

1. Davis D (2006) Urban Chinese homeowners as citizen-consumers. In: Garon S, Maclachlan P (eds) The ambivalent consumer. Cornell University Press, Ithaca, pp 281–299
2. Ho P (2001) Who owns China's land? Policies, property rights and deliberate institutional ambiguity. China Q 166:394–421
3. Huang Y (2012) Low-income housing in Chinese cities: policies and practices. China Q 212:941–964
4. Ley D, Teo SY (2014) Gentrification in Hong Kong? Epistemology versus ontology. Int J Urban Reg Res 38(4):1286–1303
5. Lin GCS (2015) The redevelopment of China's construction land: practising land property rights in cities through renewals. China Q 224:865–887

6. Lin GCS, Ho SPS (2005) The state, land system, and land development processes in contemporary, China. Ann Assoc Am Geogr 95(2):411–436
7. Pow CP (2007) Constructing a new private order: gated communities and the privatization of urban life in post-reform Shanghai. Soc Cult Geogr 8(6):813–833
8. Pow CP, Kong L (2007) Marketing the Chinese dream home: gated communities and representations of the good life in (post-) socialist Shanghai. Urban Geogr 28(2):129–159
9. Shin HB (2016) Economic transition and speculative urbanisation in China: gentrification versus dispossession. Urban Stud 53(3):471–489
10. Shin HB, Lees L, López-Morales E (2016) Introduction: locating gentrification in the Global East. Urban Stud 53(3):455–470
11. Tomba L (2005) Residential space and collective interest formation in Beijing's housing disputes. China Q 184(1):934–951
12. Walder AG (1983) Organised dependency and cultures of authority in Chinese industry. J Asian Stud 43(1):51–76
13. Walder AG (1989) Social change in post-revolution China. Ann Rev Sociol 15:405–424
14. Weber M (2009) From Max Weber: essays in sociology. In: Gerth HH, Wright Mills C (trans). Routledge, New York
15. Zhang L (2010) In search of paradise: middle-class living in a Chinese metropolis. Cornell University Press, New York

Chapter 8
Confrontation

As gentrification occurs, disagreements and resistance emerge as the result of varying motivations among the discontented. Both the media and academics have been attentive to property activists in China, particularly to the phenomenon of nail households (*dingzihu,* 钉子户), usually experienced among homeowners in urban redevelopment [19, 20, 32, 39]. "Nail householders stubbornly refuse to vacate their houses, hindering the progress of urban development projects like nails that stick out and are difficult to remove" [32, p. 1167]. Meanwhile, although residents have submitted numerous petitions, the resident resistance movements regarding urban redevelopment are kept quiet by the state. The confrontations, individually and collectively, have usually run into a series of bureaucratic holdups, only a few have gained a certain degree of success [20, 37, 40]. This study also found that not all of the challenges have come from those who contest the rationale of gentrification per se. For instance, disputes have frequently occurred because of the unclear ownership and leasehold of state- and collective-provided properties. Family members might also compete for property rights and compensation. Still, more than a few grievances have been directed at the fundamental inconsistency of payment standards and the nontransparency of policy implementation at the local level. Such discontent might be relatively easy to address through reconciliation between family members and social groupings or through an adjustment of the compensation amount.

Instead of elaborating on all of these gentrification dispute types, this chapter concentrates on those property and cultural activism types that substantially question the legitimacy of gentrification and dispossession per se. It investigates the reasons for activism to uncover the meanings and social outcomes of gentrification and displacement experienced by property activists in comparison with other working-class groups. Then, it traces the process through which antagonism is restrained (but perhaps de facto endures after the project) to highlight the intricacy of state-society relations in China. This study identifies three activist groups, a majority

Parts of the findings from this chapter have been published in China Urban Studies: Vol. 9 (Y. Ning (Ed.), Beijing: China Science Publishing, ISBN: 9787030503848 (pp. 197–220)).

© Springer Nature Singapore Pte Ltd. 2022
Q. Yang, *Gentrification in Chinese Cities*, Urban Sustainability,
https://doi.org/10.1007/978-981-19-2286-2_8

of whom are private homeowners in old, inner-city neighborhoods. Although these homeowners have been offered compensation consistent with the so-called equivalent value of their former properties, their activism is provoked by the isolation of distinctive housing consumers and the demolition of self-built housing in the city. The first group comprises the owner-occupants of courtyard dwellings, particularly in the former WNA neighborhood. Most of the buildings in the WNA area were confiscated by the government and shifted to public housing. The buildings that remained in individual possession eventually became potential sources of activism by the current homeowners. The second group consists of residents who purchased commercial apartments developed by private developers during the early stage of land and housing marketization. The last group includes homeowners of self-built shelters in the inner-city neighborhoods, which often lacks legal permit for urban land use and construction.

8.1 Property and Cultural Activism

In general, the private homeowners showed greater sensitivity to property rights and onsite living rights than did the tenants and subsidized owners. Hsing [20] highlighted the historical context that could incite the homeowners' awareness of property protection at the time of redevelopment. The private owners of historic dwellings may have surrendered individual property rights and benefits to the state on many occasions, either based on the system of Mao's anti-capitalism or, later, state-led capitalism. The informants clearly distinguished their status in the redevelopment project by asserting a violation of property rights. In contrast, the tenants and subsidized owners could benefit more from the government's compensation policies.

> From the perspective of the residents living in public housing, they think the government is right. That is, they believe it is conforming to their benefits, by which they mean the benefits of the majority. However, for us, in effect, this is illegal. We are private homeowners. (R16, cash-compensated resident)

The private homeowners not only paid more for their owner-occupancy arrangements but also maintained greater consistency between their lifestyles and residential choices than did the subsidized residents. In the old CJA neighborhood, the local officials and developers incorporated three commercial apartments into the redevelopment area without informing the residents (Fig. 6.1b). The commercial apartments were built in 2002. They are multistoried and located alongside open streets and lanes (Fig. 6.1b). Rooms in these apartment types are often more spacious than are rooms in the high-rise apartments. Buildings are displayed alongside open streets and lanes; thus, the neighborhood has inherited the spatial form of neighborhoods from Mao's China (see Gaubatz [16, 17]). Originally, the three buildings were not subject to the so-called "dilapidated housing" category. However, the Chengdu Municipal Government issued a concession to state actors and developers participating in the North

Redevelopment Program to readjust the land-use rights on a larger scale (3000–6000 m²) in this area than in the declining areas [2]. The three commodity-housing buildings were included based on the extension of the land-use rights and land redevelopment scale. A middle-aged man who was in a self-employed business in the CJA district for more than ten years purchased a commodity housing unit only one year before the establishment of the project. The man indicated that, although he clearly knew the old buildings in the CJA neighborhood would be demolished soon, he never realized that the three buildings were included in the project (R95).

The historic neighborhood of the Wide and Narrow Alley (WNA), which was redeveloped for commercial use, previously featured *hutong* (alley or narrow street) neighborhoods and courtyard dwellings that were constructed in late Imperial China (Fig. 6.1c). Neighborhood relationships constituted a key distinguishing factor of lifestyles in these types of old neighborhoods from those in the gated communities.

From 2003 to 2005, the first phase of the commercial redevelopment of the WNA area affected three streets and their adjacent dwellings (*hutong*, 胡同). The project involved completely offsite resettlement. One year after the revitalization of the WNA area began, 300 of the 891 original households rejected the offer to move [22]. Most of these households were homeowners. According to a social survey conducted by the Urban Planning and Architecture School at Chongqing University in 2005 [22], 70% of the informants demanded housing improvements without relocation, another 20% of the informants simply desired to stay put. The common reasons why these households did not want to move included their communal lifestyle, neighborhood relationships and place attachment. Fieldwork in this study also showed that the residents in the historic neighborhoods attached more importance to cultural identity than did those in the *danwei* compounds. When asked about the meaning of the old dwellings and neighborhoods, the indigenous residents specified that the notions of *zai* (residence) and *hutong* (neighborhood street) represent the nucleus of Chinese mass culture, which draws its roots from Confucianism (R77). The spatial pattern—characterized by dwellings, inward courtyards and narrow streets—allowed *hutong* neighborhoods to be a purely residential space, while according to residents, the introduction of retail stores to the streets and projects to widen the roads have ultimately invaded the residential space.

> *Zai* lived in by the masses reflect a modest aesthetic, similar to mansions that accommodate government officials and symbolize an elegant aesthetic. *Yuan* (inward courtyard) and *hutong* present intimate social relations in mass neighborhoods. However, since the road was widened and retail shops were installed, the ties have been completely broken. Previously, we could easily go to neighbors' homes and chat with each other. (R77, resident relocated offsite)

Compared with Chengdu, urban redevelopment in large cities, such as Shanghai and Beijing, which have a greater volume of architectural heritage, may face further cultural conflicts. In reference to the Nanluoguxiang renovations in Beijing, Shin [31] found that, although the residents supported the state's views of cultural conservation for the renewal project, they were less confident in the methods used and results attained. This finding is attributable to the fact that so-called conservation has been characterized by, on the one hand, large-scale commercialization and residential relocation and, on the other hand, the reconstruction of large structures while retaining only a small number of authentic originals. Based on these concerns among local residents in historical sites, property activists have highlighted the nature of the redevelopment project as involving a recreation of the urban imagery and over commercialization led by local officials.

Thus, for private property owners, property requisition and compensation are not relevant to a change in tenure type; rather, they connote a forced transition of the housing character. The project is not a welfare project, and it does not necessarily reflect the expropriation of benefits by private developers through land capitalization and housing commodification. Rather, it is undertaken by local government and officials to seek political achievements and a historically themed tourist place through urban reconstruction.[1] Unlike residents who depend on housing welfare, private property owners are more likely to recognize the irrationality of property acquisition and the emotional damage inflicted by a forced lifestyle change.

> When [the demolition office staff] asked my opinion, I told them I did not want the housing to be demolished. Our house does not need redevelopment. It was part of the commodity housing built in 2002. This is commodity housing, while the other so-called private housing in the CJA neighborhood is in effect subsidized housing. My living room is as large as 38 m². There are still two big bedrooms. Each is more than 16 m². The bedroom of the current high-rise apartments is regularly approximately 10 m². I also do not want to engage in property speculation. We only have this house. It is not an illegal house; the building has no question marks. We have completed two certificates (land-use and property ownership). Is it not legal for us to reject demolition? What is the problem with us continuing to live here? (R16, cash-compensated resident)

Following the redevelopment of the WNA and CJA neighborhoods, most of the activists ultimately moved out, as they either acquiesced to the higher compensation amount or submitted to coercion. For the residents who had a weak awareness of property rights, the negotiations with the state have easily interwoven and blurred with the request for residential rights onsite, the preference for a distinctive lifestyle and an existing tendency for profit-seeking. Thus, the claim for economic compensation easily becomes the only tool that private owners can use to achieve high levels of "equality." However, those with a higher legal consciousness would argue for the protection of consolidated private property rights. For instance, they may argue for in-kind compensation located at an adjacent location and for similar housing and neighborhood styles as those they previously had.

> They explained that this is based on urban planning and policies. I told them they were deceiving me. I work in the construction industry. I know urban planning. The three buildings are located alongside a road near the border of this land. It will not impact your use of the land. Planning is subject to change when it is not fixed. However, later I also said, "If you forced me to move and were concerned with the responsibility of your sector, the performance of your government, to satisfy you, I had only one demand. Please find me a similar house with a similar dwelling design, a similar location and a similar community. It should be multistory, not high rise." (R16, cash-compensated resident)

---

[1] Interviews with R16, 77, 80, 87, and 96.

Some of the activists, including both those who have moved out of their former neighborhoods and those who have stayed put, have involved themselves in an extremely laborious and distressing confrontation process with the government and developers to condemn the unfair treatment in the redevelopment and resettlement process or to continually protect their housing. After being forced to move out, Shun, one of the commercial apartment owners in the CJA neighborhood, involved himself in administrative lawsuits against the District Public Security Department and the District Government for more than two years. However, the absence of a legally permitted and publicly acknowledged channel of protection of private property rights has largely reduced the efficiency of resistance (see [30, 36]). For instance, having been forced to sign an agreement for demolition and removal, Shun could only charge the District Public Security Department with his brutal treatment instead of directly accusing the District Government of perpetrating illegal demolition and removal (R95).

Mr. Yang is the only activist who preserved his dwelling in the WNA neighborhood, but his family no longer lives there because the area has been substantially commercialized (Fig. 8.1). Yang is a retired professor of art at a local university. He is 70 years old and Manchu. Yang has devoted his whole life to cultural undertakings, particularly Manchu cultural preservation. Yang and his family have lived in the WNA area for more than half a century; his wife inherited the house from her mother before the liberation of China. To date, Yang's resistance to allowing his property to be demolished has lasted for seven years and is currently at a stalemate. Even several years after the street was redeveloped for commercial use, the developer could still demolish Yang's courtyard wall at any time and occupy part of his dwelling. Moreover, the city residents seldom care about tradition and culture, and they refuse to align themselves with those resisting the takeover, which has massively disappointed the activists. Ironically, as Yang stated, "It is owing to my acquaintance with foreign friends that the house has been preserved thus far" (R77). According to Yang, one of the cultural activists from the WNA area finally emigrated from China as a result of the detrimental effects of confrontations with the government and developers.

As we walked around the shopping street of the WNA area, Yang pointed out to me individually those apparently "old" structures that are actually fake, providing a

Fig. 8.1 Wide and narrow Alley before (2004) and after commercial redevelopment (2016). *Source* Photos taken by the author and provided by a local resident in WNA

striking contrast to tour guides who told fabricated stories to tourists. Passing through the door of Yang's property, I saw that the yard was divided into two parts by a wall. One half has been redeveloped as a theatre, and the other half was being reused by Yang as a teahouse. In the teahouse yard, Yang has reserved the tiles and bricks from the old structures from the demolition. In front of the door, two young waitresses held a shop sign and received guests to the theatre, while the small teahouse was largely ignored by the tourists. The shop sign noted the name of the yard, and it was made and written by Yang in a kind of Chinese calligraphy.

8.2 Claiming Legitimacy for Informality

Another source of property activism derives from clearances for informal shelters built by citizens without legal permission for land use and construction. The removal of informal structures and their replacement with formal, governable landscapes threatens an alternative urbanism—economic activities, lifestyles and social relations in informal urban sectors. In a central debate in studies on informal land uses, informal sectors are deemed either as having a historical legacy due to their underdevelopment and marginalization by the formal economy (i.e., the traditional viewpoint) or as being a permanent part of urban society and the economy [1]. The latter viewpoint stresses that urban informality is a part of the diversity of urbanization in developing societies [1, 29]. Self-help housing is necessary to alleviate the economic poverty and housing difficulties of the urban poor, especially when available housing is in short supply in the collective and private housing market in the city [26, 34]. The dichotomy in the understanding of urban informality leads to disputes on the methods for tackling questions regarding informality. The former position may advocate incorporation into the formal system to enable actors in informal sectors to hold legal property rights and participate in the formal economy [11–13]. The latter position challenges the efficacy of formalization through simply legal entitlement because such a straightforward method might ignore the established economic patterns and practices of the informal sectors and the obstacles they face in participating in the formal sectors. Such a method may engender other difficulties for informal inhabitants because it unwittingly results in the formal system being the exclusive operational mechanism [18, 21, 27–29].

In China, urban informality has attracted considerable attention from researchers of urban villages (following strong criticism of the rapid urbanization and land acquisition in China). Wu et al. [38] revealed that informal settlements in the city were first developed due to the shortage of low-income rental housing. More importantly, many of the so-called illegal structures essentially exist because of the dual urban–rural land system and the failure of property rights registration and redistribution as urban areas expanded to rural districts. Nonetheless, the informal structures identified as "illegal" have been erased, and the areas have then been reconstructed as master-planned communities. Displaced residents have been entitled to compensation only for construction, materials and decoration expenses.

As a result, the most severe resistance could be stirred up by homeowners of the titled informal structures in urban areas, as well villages in the city. The reason for their resistance is not simply the desire to receive compensation equal to that based on complete property rights, or even a claim on property rights. Rather, it is the result of insecurity engendered from changes to economic functions and livelihoods achieved only through informal structures. For instance, Fuzhen Tang, the subject of a self-immolation event in Chengdu in 2009 that caused nationwide concern, was a successful, affluent entrepreneur in a family business of garment production and wholesaling. Her demolished three-story housing building was more than 2000 m^2 and was used as a garment-processing factory, an office and a residence [7]. Approximately two decades ago, Tang and her husband worked in Chengdu in wholesale clothing. The village committee invited them to return to Tang's rural hometown and invest in village industries. They acquired the land-use agreements from the village committee and built the factory and house, but without legally issued permits for land use and construction [14]. Disregarding the significance of the building for Tang's business, the local officials judged Tang's self-burning behavior as violent resistance to the law. Tang had demanded cash compensation as high as eight million Chinese yuan (approximately US$1,230,769), compared with the two million Chinese yuan (approximately US$307,692) in compensation for the construction cost promised by the local government [25].

The conditions of the self-built homeowners in *danwei* compounds are slightly different from those of the landlords on previously rural land. While the landlords have mostly acquired the land from village committees [10, 32, 38], the urban homeowners have usually built informal structures in public spaces. Therefore, the size of the independent, multistory dwellings in urban villages is much larger than the additional structures in old inner-city neighborhoods. In the old CJA neighborhood, ownership claims on the informal structures are mostly overlaid with claims by homeowners of formally subsidized housing or by public tenants. The informal structures were used by the locals to either accommodate their growing families or set up personal businesses. Vendors used informal structures to sell daily necessities and food to their neighbors at relatively lower prices than those in the formal market. The small stalls brought the locals and migrant vendors significant income by means of self-employed businesses or rent. The local residents might have used the rental incomes thus generated to pay for a larger rental apartment to adapt to an increase in family size.

Mrs. Yu, a subtenant in CJA, built two booths in the informal market in 2002. After the first few years selling rice wine, she leased them to another vendor, who maintained the rice wine business. It has earned her more than 4000 Chinese yuan (approximately US$615) in rent every month. However, the initial tenant of the apartment where she lives with her son is her former husband. Thus, she now has no direct right to either renew the tenancy of the public housing or purchase the rental housing to exchange it for private housing. In the national media, she stated her demands:

I do not have any residence. I only have the two shop fronts. They are the source of my son's school fees. If you are going to demolish here, I will insist on compensation for my shop fronts. Moreover, I have been selling the rice wine here for up to ten years. It is so popular here. Actually, rice wine should be seen as part of the "intangible cultural heritage" in Chengdu. You should protect it. (Resident interview by China Central Television, 2012)

The shanty homeowners primarily demand familial privacy. They constructed shelters in public areas, such as staircases or verandas, thereby creating spatial separation between different generations in the home and avoiding embarrassment and inconvenience. Because the in-kind compensation does not include the shelter areas, they may still be living in crowded spaces. Meanwhile, the residents have argued that they will lose the opportunity to build any additional structures for future separation. Even worse, such limited living spaces will continue to affect the childbearing of their children's families.

Now I am living in the staircase. ... I need an apartment with two bedrooms only so I will not impact my son and his wife. Additionally, they can consider having a baby. I have no choice. Because I am retired, I cannot gain the basic subsistence allowance. Without it, I cannot apply for low-rent public housing. The only place I can live is in the exchanged apartment. If I chose monetary compensation, I could only gain approximately 200,000 Chinese yuan and that actually was not enough even for a down payment on the market housing now. (Resident interview by China Central Television, 2012)

The current literature has mostly focused on the harm to homeowners done by the informal structures because they depress the rental income generated by multistory redevelopment [9, 20, 32]. The above analyses indicate that the informal structures signify the work and lifestyles of their inhabitants, which were established based on their socioeconomic positions in the city and institutional environments. By examining the correlation between urban informality and the socioeconomic performance of inhabitants, Li and Wu [23] and Webster et al. [35] proved that residential satisfaction is not necessarily low in informal settlements, indeed, residents could take advantage of the ambiguity of the property rights of the informal structures to overcome poverty. The study found that homeowner resistance might not even be a result of poverty (see also [29]). The urban villagers may have accumulated substantial wealth from economic activities in the informal sectors. State-facilitated gentrification in China, replacing spontaneous spatial formation with spatial regulation and commodification, is associated with not only the issue of property rights but also the reconfiguration of the economic activities and lifestyles of in situ social groups. The strict control of urban informality has led Chinese cities to become a predominantly spectacular landscape with a relative lack of visible slums. As Wu et al. [38] argued, the redevelopment of urban villages has been based on the idea of erasing informal spaces and recreating "governable spaces through formal land development" (p. 1919).

8.3 Mass Mobilization and Consensus-Building

8.3.1 Preliminary Agreement Policy for Removal

In the face of conflicts and confrontations, the local government has employed multiple social governance tools. The first social governance tactic derives from a policy innovation at the local level. The local government has combined two national policies of property conveyance and compensation to enhance the efficiency of consensus-building among the majority of residents and maintain state actors' power in wielding instruments of forcible eviction against property activists.

Before 2011, according to the central policies of *Housing Demolition and Compensation*, private companies were allowed to charge for property demolition in an urban redevelopment project. Private participants had to apply for a demolition permit from the local governments to be authorized to undertake the housing acquisition, consensus-building and compensation process. This property demolition led by private companies was defined as a *market* behavior of property acquisition based on the principle of free market negotiation. Based on the policy context, the local government initiated a technical approach called "preliminary agreements on demolition and removal" (*moni chaiqian xieyi*, 模拟拆迁协议) to arrange consensus-building regarding demolition and removal among residents. Preliminary agreements on demolition and removal had to be entered into before a redevelopment project could officially be launched. The demolition party had to gain approval before a certain deadline from a certain number of residents with respect to demolition and removal based on a series of predetermined conditions regarding demolition, removal and compensation. Only when 95% of all the households approved the project was it officially established[2]; various sources then started to fund the project [3, 6, 8]. The other 5% of the households—the so-called nail households—had to be forced to accept property demolition and removal based on the demolition policy of administrative enforcement (i.e., by the government or local housing authorities). If the required proportion of original homeowners did not approve during the process of obtaining preliminary agreements on demolition and removal, the project was postponed (O2, official from a task force of the North Chengdu Redevelopment Program).

Since 2011, however, the State Council has redefined property conveyance as an *administrative* behavior that is essentially justified by state authority in which only the government is accountable. Accordingly, the "property demolition" (*fangwuchaiqian*, 房屋拆迁) discourse in the central policies has been replaced with "property expropriation" (*fangwuzhengshou*, 房屋征收; [33]). Through the latter, the central government has also prohibited local governments from granting demolition permits to private companies and has terminated the government execution of forcible demolition. Rather, demolition enforcement responsibility has been transferred to judiciary departments. Under the property expropriation policy, the launch

---

[2] The proportion is based on the 2012 norm. It was 85% based on the 2008 norm.

of an urban redevelopment project and subsequent property acquisitions must thus be submitted to extended administrative procedures.

Nevertheless, compared with the negotiable principle of property acquisition led by developers, local governments deem the administrative procedures lacking in practicability and inefficient. Local governments have thus continued to adopt a mode of property conveyance based on market principles. However, to avoid direct conflicts with the central policies, the local policies of urban redevelopment have followed the regulations of the central policies forbidding the approval of demolition permits to third parties and the use of forcible demolition by the government to ensure that the government is accountable for demolition. Additionally, the local policies have shifted their language use from "property demolition" to "property expropriation" [4, 8]. Since 2011, there have been essentially no cases of urban redevelopment based on the mode of administrative expropriation (O22). This "compromise" between the two policy rounds has allowed local governments to bypass complicated official rules and has continued to promote the highly efficient mode of property acquisition and consensus building. Local governments have thus reserved the previous method of obtaining "preliminary agreements on demolition and removal" in recent projects. However, they changed the phrase to "preliminary agreements on *removal" (moni banqian,* 模拟搬迁协议) to avoid using the term "demolition."

Nevertheless, by evading the administrative acquisition procedures, the demolition parties have lost any coercive force supported by either the government or judiciary departments. After 95% of the residents have signed the agreements, the demolition parties have no power to evict the 5% who oppose the demolition. In this case, the local policies require complete residential approval in the process of obtaining preliminary agreements [5]. This approach, to a certain degree, restrains the occurrence of brutal behavior while increasing the use of soft governance forms in consensus-building. However, the approach cannot fundamentally stop property activism. In nail household cases, local governments have explored a new norm of property acquisition by suggesting a conflation of the modes of "market acquisition" and "administrative expropriation" for one redevelopment project. This policy is called "transferring from the stage of preliminary agreements on removal to targeted 'property expropriation'" (*dingxiangzhegnshou,* 定向征收; O22). The phrase denotes a relegitimation of forcible eviction by returning to judiciary enforcement when coercive instruments are needed to acquire particular properties.

The policy of preliminary agreements on removal greatly encouraged local officials to convince residents to agree during the preliminary stages as efficiently as possible. In the CJA project, the demolition sector affiliated with the Beixin Company was directly responsible for negotiating with residents. Profit concessions on the lost-housing compensation of the current residents was a common practice used by the government to encourage the existing residents to accept the plan (see also [24]), such as early resettlement rewards by any names, increases in the minimum compensation area and discounts for buying additional areas of the resettlement housing. Nevertheless, the method of obtaining preliminary agreements on demolition and removal greatly encouraged willing movers to participate in consensus-building.

Willing movers and speculators worried about delays to the project and sponta-neously worked to provide conflict mediation and persuade holdouts. In principle, the method also shortened the redevelopment project cycle and brought about the greatest efficiency of fund usage.

8.3.2 Autonomous Redevelopment Committee in Action

However, the questions hindering the progress of the project were complex and mani-fold. These questions concerned issues not only among the government and commu-nities but also among social groups and family members. For instance, the resi-dents behaved radically toward malpractice and corruption by party members at the local level. They challenged the implementation consistency of policies and norms, especially considering the higher compensation for private owners of commercial apartments.Above all, the anticipation of substantially improved living conditions caused widespread dissatisfaction among residents regarding the minimum area of the resettlement apartment per household and its interior design. Additionally, the mismanagement of lease contracts for public housing caused bitter feuds among family members regarding the identity of the initial public-housing tenant.

Given the intricacy of the social conflicts in the project, a so-called mode of "autonomous redevelopment by residents" (*jumin zizhi gaizao*) was proposed by the district government at the outset to ease consensus-building throughout the entire project (see also [15]). An autonomous redevelopment committee was encouraged by the government and established. Twenty-one committee members from among the residents in CJA were elected through open voting by the CJA residents. The members selected were almost all of the residents in public/subsidized housing who were living in the worst housing conditions. They were considered to be representative of the living conditions and needs of most of the families in the community.

This committee had the following functions: first, to understand resident appeals; second, to consult and negotiate with the government on these needs; and finally, to mobilize as many residents as possible in support of relocation through persua-sion, argument or emotional resonance. In practice, the committee members led the process of social investigation, organized several rounds of mobilization meetings, arbitrated among different parties, and impacted the plan-making for resettlement. Most frequently, they engaged in conflict mediation among family members. For example, in the face of family disputes over housing entitlement, the committee members would persuade any of various family members daily to relinquish housing rights and to consider another way of housing themselves.

The most important contribution of the committee for the majority of the CJA residents was to achieve the claim for at least two bedrooms for each household after compensation. This step was crucial for reaching full agreement from all of the public tenants and subsidized owners. Against the advice of the committee members, the local government adjusted the minimum onsite compensation area for one household previously living in a 16 m^2 unit from 48 to 58 m^2. The request caused the district government and Beixin Company to organize several rounds of urban design for the onsite resettlement neighborhood to rehouse as many residents as possible. As a result, while ensuring the social and economic benefits of the project, the Jinniu District government and Beixin Company endeavored to convince the Chengdu Municipal Planning Institute to expand the plot ratio. Mixed with new commercial and resettlement housing, the new CJA housing area ultimately achieved a plot ratio of more than 6, twice the local criterion (O16 and 17, officials from the Urban–Rural Housing Department, and R3 and 4, residents relocated on site).

To thwart any intense opposition, the local governments continued to resort to the Autonomous Redevelopment Committee. Frequently, the committee members offered earnest advice to the opponents by portraying their vision of the beautiful life that they would live in the newly built resettlement community compared with their lives in the degenerated *danwei* compound. The infrastructure conditions, the housing styles, and the social environments of the old and new communities were compared side by side; the lifestyle in the old neighborhood was described as 'backward' and thus was bound to be eliminated. In some cases, the committee members also suggested that the government reward them for attaining the agreements. In other cases, committee members tended to remind those who expected even higher compensation prices for the old housing to not "spoil the ship for a ha'p'orth of tar".

The persuasion methods accentuated the poor reputation of the old neighborhood and, rhetorically, enhanced the confidence of the residents that the resettlement community would bring about a life worth living. This social impact helped to expand the residents' suffering from the government's relocation plan and further justified the tendency of the local governments to subordinate the will of the minority for the sake of the majority. Ultimately, the negotiations between the government and residents was, to a certain degree, internalized to become a negotiation among the residents.

As the project was proceeding, some mild opponents increasingly chose to accept the rehousing scheme. Within only four months (from March 9 to June 17, 2013), the CJA redevelopment project achieved approval from all of the subsidized owners and public tenants and another 101 households in commodity housing for demolition and removal. However, 27 households in commodity housing remained holdouts and were called "nail households" by locals—property activists who "stubbornly refuse to vacate their houses, hindering the progress of urban development projects, like nails that stick out and are difficult to remove" [32, p. 1167]. The Autonomous Redevelopment Committee then proposed an extension to the government so it could obtain the preliminary agreements regarding removal.

8.3.3 Enforcement

Unfortunately, among the 27 households, coercive measures were launched by the administrative and judicial systems in extreme cases. First, through deception and menace, the demolition office, affiliated with the district government, urged the opponents to sign agreements regarding demolition and removal. One owner-occupant of a commercial apartment described his experience:

> At 6 pm, the demolition office sent people to my house. They did not allow me to leave and forced me to sign the contract. Later, the Autonomous Redevelopment Committee also came. They surrounded me and prohibited me from going to the toilet and drinking water. Later, several people in hoodies entered, with beer in their hands. People from the demolition office said to me that they had taken drugs. They were warning me that they [the people in hoodies] might do something unexpected. (R46, cash-compensated resident)

Notably, some of the committee members and other energetic, willing movers might have also involved themselves in those actions since the resisters often chose to leave the community temporarily to escape disturbances from the demolition office. Once the resisters appeared in the community, the committee members would prevent them from leaving and constantly ask for their consent. Thus far, the bond of neighborly and collegial relations between the willing and involuntary movers was fractured due to the incompatible housing demands. A new connection was shaped among the willing movers of a preference for the new housing style featuring modernity, possessions and private lives, ultimately arousing collective antagonistic behaviors against their opponents.

> R12: The demolition success is indeed the result of our efforts. At that time, the community situation was truly bad. However, I think we also offended many neighbors. However, it should be understandable because we were working for the good of the majority. I did not gain any benefit by doing this job (mediating conflicts). We stayed up all seven nights.
>
> Researcher: Why did you stay up all seven nights?
>
> R12: They did not agree to sign (the agreement). We just kept watch on them and asked them to sign. (R12, resident relocated offsite)

A month later, fifteen of the 27 "nail households" signed the agreements. The demolition party and the government then decided to make a temporary adjustment to the redevelopment scope by excluding the building of the last 12 homeowner areas to end the stage of the preliminary agreements on removal. The project was officially launched on July 15, 2013.

However, during the next year, the 12 holdout households were forced out by the judiciary system. Based on the local policies of housing expropriation, after being informed by the government, the households that refused to relocate were allowed to propose an administrative review by the municipal government of the property expropriation decision and be proposed compensation in 60 days or file a lawsuit directly against the government. Otherwise, they were forced to accept property expropriation by judiciary enforcement. From the end of 2013 to April 2014, the district court deliberated on the compulsory execution of the demolition of the last building, in which the 12 remaining households lived. On 27 April 2014, the last household was removed from the CJA neighborhood.

After the CJA neighborhood was redeveloped, most of the activists either moved into resettlement communities or received cash compensation; a few involved themselves in the extremely laborious and distressing process of condemning the unfair redevelopment and resettlement treatment from the government. Despite a sense of displacement, these activists have ultimately become the new type of urban citizen by entering gated communities and moving into high-rise apartments. The increasing disappearance of cultural distinctiveness and the convergence to a standardized housing style are expected to further weaken any defenders of those distinctive lifestyles and local identities.

References

1. AlSayyad N (2004) Urbanism as a "new" way of life. In: Roy A, AlSayyad N (eds) Urban informality: transnational perspectives from the Middle East, South Asia and Latin America. Lexington Books, Lanham, MD, pp 7–30
2. Bureau of Land and Resource of Chengdu (2012, May) Chengdu beigai tudi liyong shishi xize (shixing) [Regulations on the implementation of the North City Redevelopment programme in Chengdu (trial)], No. 116. Retrieved 11 July 2015, from: http://www.cdlr.gov.cn/detail.aspx?id=37808
3. Bureau of Urban-Rural Housing of Chengdu (2008, Oct) *Guanyu zhongxin chengqu chengshi fangwu moni chaiqian gongzuo de zhidao yijian (shixing) de tongzhi* [Notification on "The opinions of the implementation of the preliminary agreements on demolition and removal in the central city"(trial)], No. 146. Retrieved 6 Dec 2014, from: http://www.cdfgj.gov.cn/News/ShowInfo.aspx?ArticleGuid=6B42F92A-C019-4DD4-A10F-7E9358D0FC8A
4. Bureau of Urban-Rural Housing of Chengdu (2012, Dec) *Guanyu guoyou tudishang fangwu zhengshou yu buchang gongzuo zhong jige juti wenti de tongzhi* (Notification on several issues about housing conveyance on state-owned land and compensation). No. 131. Retrieved 22 Feb 2013, from: http://gk.chengdu.gov.cn/govInfoPub/detail.action?id=72493&tn=6
5. Bureau of Urban-Rural Housing of Chengdu (2012, May) *Guanyu guanche zhixing "chengdushi renmin zhengfu bangongting guanyu jinyibu cujin beicheng gaizao youguan zhengce de yijian" de tongzhi* (Notification on implementing the "Opinions on the policies for the promotion of North City Redevelopment" from Chengdu Municipal government), No. 27. Retrieved 22 Feb 2013, from: http://www.cdfgj.gov.cn/cqgl/ShowInfo.aspx?ArticleGuid=410c9b1c-ce3b-41bf-929e-e43b734c0d62
6. Bureau of Urban-Rural Housing of Chengdu (2012, Nov) *Guanyu zai fangwu zhengshou zhong zuohao moni banqian gongzuo de zhidao yijian* (Opinions on implementing preliminary agreements on removal in housing conveyance), No. 36. Retrieved 8 Sept 2015, from: http://www.cdfgj.gov.cn/CQGL/ShowInfo.aspx?ArticleGuid=a449defe-e4f8-4784-a893-ca1de5d251dc
7. Chen R (2010, Jan 25) A woman burns. New York Times. Retrieved 13 Aug 2015, from: http://www.nytimes.com/2010/01/26/opinion/26iht-edcohen.html
8. Chengdu Municipal Government (2013, Dec) Guanyu jinyibu tuijin zhongxin chengqu jiucheng gaizao guifan fangwu zhengshou yu buchang xingwei de tongzhi (Notification on promoting urban renewal in the central city and regulating housing conveyance and compensation), No. 57. Retrieved 10 July 2015, from: http://www.chengdu.gov.cn/wenjian/detail.jsp?id=ewAnYyOFWYfcUHepUf0E
9. Chung H, Zhou S-H (2011) Planning for plural groups? Villages-in-the-city redevelopment in Guangzhou city China. Int Plan Stud 16(4):333–353
10. Chung H (2010) Building an image of villages-in-the-city: a clarification of China's distinctive urban spaces. Int J Urban Reg Res 34:421–437
11. De Soto H (1989) The other path: the invisible revolution in the Third World. I. B. Taurus, London
12. De Soto H (2000) The mystery of capital: why capitalism triumphs in the West and fails everywhere else. Basic Books, New York

13. Deininger K, Binswanger H (1999) The evolution of the World Bank's land policy: principles, experience, and future challenges. World Bank Res Observer 14(2):247–276

14. Deng A (2011) Dousing the flames: the Tang Fu Zhan self-immolation incident and urban land reform in the People's Republic of China. Southern California Interdisc Law J 20(3):585–617

15. Deng YH (2017) Autonomous redevelopment: moving the masses to remove the nail households. Modern China 43(5):494–522

16. Friedmann J (2007) Reflections on place and place-making in the cities of China. Int J Urban Reg Res 31(2):257–279

17. Gaubatz P (1999) Understanding Chinese urban form: contexts for interpreting continuity and change. Built Environ 24(4):251–270

18. Gilbert A (2002) On the mystery of capital and the myths of Hernando de Soto—what difference does legal title make? Int Dev Plan Rev 24(1):1–19

19. He S (2012) Two waves of gentrification and emerging rights issues in Guangzhou, China. Environ Plan A 44:2817–2833

20. Hsing Y (2010) The great urban transformation: politics of land and property in China. Oxford University Press, New York

21. Kagawa A, Turkstra J (2002) The process of urban land tenure formalisation in Peru. In: Payne G (ed) Land, rights, and innovation: improving tenure security for the urban poor. ITDG Publishing, London, pp 57–75

22. Li XB, Wang P (2007) Gongzong canyu yu chengshi lishi jiequ gengxin de tujing: Yi Chengdu kuanzhai xiangzi weili (Social participation in the renewal of historic area: a case of the wide and narrow Alley in Chengdu). J Xihua Univ Philos Soc Sci 2:109–110

23. Li Z, Wu F (2013) Residential satisfaction in China's informal settlements: a case study of Beijing, Shanghai, and Guangzhou. Urban Geogr 34(7):923–949

24. Lin GCS (2015) The redevelopment of China's construction land: practising land property rights in cities through renewals. China Q 224:865–887

25. Liu L (2009, Nov 26) Chengdu yi nyuqiyejia yin chaiqian louding zifen (An entrepreneur in Chengdu burns due to housing demolition). New Hunan Newspaper. Retrieved 13 Aug 2015, from: http://www.boxun.com/news/gb/china/2009/11/200911262243.shtml

26. Mooya MM, Cloete CE (2007) Informal urban property markets and poverty alleviation: a conceptual framework. Urban Stud 44(1):147–165

27. Razzaz O (1997) Legality and stability in land and housing markets. Land Lines 9(3):1–4

28. Roy A (2003) Paradigms of propertied citizenship: transnational techniques of analysis. Urban Aff Rev 38(4):463–491

29. Roy A (2005) Urban informality: toward an epistemology of planning. J Am Plann Assoc 71(2):147–158

30. Shih M (2010) The evolving law of disputed relocation: constructing inner-city renewal practices in Shanghai, 1990–2005. Int J Urban Reg Res 34(2):350–364

31. Shin HB (2010) Urban conservation and revalorisation of dilapidated historic quarters: the case of Nanluoguxiang in Beijing. Cities 27:S43–S54

32. Shin HB (2013) The right to the city and critical reflections on China's property rights activism. Antipode 45(5):1167–1189

33. State Council of the People's Republic of China (2011, Jan) Guoyou tudi shang fangwu zhengshou yu buchang tiaoli (Ordinance of housing conveyance and compensation on state-owned land), No. 590. Retrieved 10 July 2015, from: http://www.gov.cn/zwgk/2011-01/21/content_1790111.htm

34. Turner JFC (1976) Housing by people. Towards autonomy in building environments. Marion Byers, London

35. Webster C, Wu F, Zhang F, Sarkar C (2016) Informality, property rights, and poverty in China's "favelas." World Dev 78:461–476

36. Wilhelm K (2004) Rethinking property rights in urban China. UCLA J Int Laws Foreign Aff 9:227–300

37. Wu F (2004) Residential relocation under market-oriented redevelopment: the process and outcomes in urban China. Geoforum 35(4):453–470

38. Wu F, Zhang F, Webster C (2013) Informality and the development and demolition of urban villages in the Chinese Peri-urban area. Urban Stud 50(10):1919–1934
39. Zhang L (2004) Forced from home: property rights, civic activism, and the politics of relocation in China. Urban Anthropol 33(2–4):247–281
40. Zhang Y, Fang K (2004) Is history repeating itself? From urban renewal in the United States to inner city redevelopment in China. J Plan Educ Res 23(3):286–298

Chapter 9
Submission

Private tenants in the old inner-city neighborhoods are almost all low-income migrants, and a large part of them are rural–urban migrants who do not have the entitlement of an urban *hukou*. The National Bureau of Statistics in China recorded that in 2006, 150 million people, or 11% of the population, were registered as peasant migrants in the city [3]. There is no data on the number of migrants or private tenants living in the redeveloped neighborhoods because the population is transient. According to the census data at the subdistrict level, which is slightly larger than the scope of the redeveloped neighborhoods under investigation, migrants accounted for 70.13% of the registered population in the census tract including the JW neighborhood, and 40.81% in the census tract including the CJA neighborhood, in 2010. The shares of migrants with a rural *hukou* in the entire registered populations were 50.21% and 20.02% for the JW and CJA neighborhoods, respectively. In both the JW and CJA neighborhoods, private tenants accounted for the largest tenure group in the 2010 data, composing 47.21% and 44.35%, respectively, of all of the tenure households in the subdistricts.[1] The Wide and Narrow Alley (WNA) has a different situation from that of the above two cases, as it originally contained historic dwellings and was redeveloped in the early 2000s. In 2000, before the WNA area was revitalized, 34.07% of the population in the census tract including the WNA neighborhood were migrants, while 12.53% were involved in agriculture (Population census, 2000).

The 2001 urban redevelopment policy ended rental regulations on both public and private rental markets. Since then, no formal policies have protected the leasing status of private tenants. As the low-income migrants cannot formally receive any compensation in the private rental market, gentrification has directly caused their eviction from the old inner-city neighborhoods.

Parts of the findings from this chapter have been published in China Urban Studies: Vol. 9 (Y. Ning (Ed.), Beijing: China Science Publishing, ISBN: 9787030503848 (pp. 197–220)).

[1] 2010 census data at the subdistrict level were collected from the five district governments in Chengdu.

© Springer Nature Singapore Pte Ltd. 2022
Q. Yang, *Gentrification in Chinese Cities*, Urban Sustainability,
https://doi.org/10.1007/978-981-19-2286-2_9

9.1   Silent Migrants

Facing this predicament, the migrants, however, showed widespread apathy about inner-city gentrification. Often, without any resistance, migrants must search for new rental housing before a redevelopment project even begins, based on either their knowledge about the project or information provided by landlords. The media also seldom bring them into the limelight, compared with the self-housing homeowners discussed in the last section. Shin [7] thus stressed that the disadvantaged migrants have been powerless in the confrontation with the government in achieving rights to the city, the current literature has typically concentrated on homeowners rather than migrant tenants regarding anti-eviction plans. During the interviews with the displaced migrants in an old neighborhood near the CJA project, our talk frequently digressed from urban redevelopment to their more general working lives in the city. They did not appear to care about their place in the process of urban redevelopment, nor were they resistant to displacement. The interview contents frustrated me, until the lengthy conversations enabled me to portray a broader picture of the migrants' lives in the city. Their obedience and passive responses remained in accordance with their perceptions regarding their housing and working status in the city, which has been cultivated by the long-term experiences of spatial eviction, sociocultural exclusion and institutional marginalization.

The first reason that the migrants are inactive is that they perceive urban society to be irrelevant. This perception echoes Solinger [9], who revealed that migrants could establish themselves outside the formal city systems. Indeed, as they have detached themselves from mainstream society, not a few of my interviewees conveyed that their status was unrelated to urban affairs. However, as this study shows, experiencing urban redevelopment and displacement has caused migrants to increasingly recognize that their livelihoods in the city are threatened. Ultimately, they displayed a more grounded understanding than even the nonmigrant urban poor. Then, in comparison with the migrant tenants' experiences of residential mobility, the residential displacement caused by urban redevelopment is hardly unique. Within the old neighborhoods, their informal status has led to high residential and business mobility. They frequently move, chiefly because of urban redevelopment but also because of job changes, the unaffordability of rental prices, conflicts with landlords and business eviction by city managers.

> I have been living in the area of Balizhuang for many years. Initially, we lived in the fourth block. When they began to demolish it, we started moving. First was the eighth block, then the ninth—one by one. Now, we live in the sixth block of Balizhuang. Usually, when we signed a contract, the landlords would tell us an estimated demolition date. Then, the contract deadline would be approximately two months before that date. We would pay a deposit of only one month's rent. If the housing was not torn down on time, then we simply extended the contract by a few months. (R56, displaced resident)

Their inactivity toward displacement represents a deeply pessimistic attitude regarding their survival conditions and living rights in the city. The migrants were usually reluctant to participate in the interviews. They declined not necessarily because of wariness about my intentions but simply because of their unwilling-ness to talk. Talking with a stranger like me would do nothing to better their life regardless of who I am, whether I work for the government or critical media, or how powerful I may be. Pun [6] argued that ideologically, the negation of rural lifestyles by the state, as the antithesis of urban modernity and consumerism, along with the official discursive practices that tend to compress the political connotation of class difference have contributed to the aphasia of the floating population.

> Researcher: Do you think demolition is unfair to you?

> R36: You lived in the landlord's house. How can you say anything? Moreover, who is talking about justice now? Where is justice? We are bullied everywhere.

> R37: Hey, the state only needs a policy to stop you from doing something. Previously, during the Cultural Revolution, they said that we were profiteers (*toujidaoba*). We thus made a mistake when we simply sold eggs. Later, they allowed farmers to enter the city, and then we came in for business. Later, Li Chuncheng (the former mayor of Chengdu) wanted to establish the Hygienic City. They thus prohibited us from setting up stalls. They said we were dirty. Now they say the place is going to be cleared away; then we should just move. Who knows what kinds of policies they will make in the near future? (R36 and 37, displaced residents)

9.2 Losing Livelihood Spaces

In addition to residential eviction, the rural–urban migrants have faced new types of social exclusion after being displaced from the old neighborhoods owing to spatial commodification and formalization in the inner city. Most significantly, the newly built mode of urban redevelopment has caused the displaced tenants to suffer from increasingly severe issues of employment displacement in the inner city. The displacees participating in this study had mostly moved from rural areas to the city during the late 1990s and later moved to the old CJA area. Currently, they live with two or three generations under one roof, with an oldest generation that was born in the 1960s. The rural–urban migrant interviewees were self-employed, or they worked in low-end service and retail jobs. In particular, many earn their living by selling agricultural products or food at the farmers' market in the old CJA area (Fig. 9.1).

In this case, the rural–urban migrants expressed their great dependence on living and working in the old neighborhoods (see also [14]). Accordingly, these tenants are highly localized and are concentrated in different types of old neighborhoods or in informal, less desirable places of the city (see also [5, 12, 13]). Shortly after the demolition of the CJA area began, the vendors moved and occupied a plot of land in an adjacent old neighborhood of the CJA area. Another farmers' market was soon formed in this old neighborhood, called Workers' Village. This Workers' Village, however, was supposed to be demolished after the CJA redevelopment project. As a consequence of the increasing formalization of commercial places, low-income migrants have faced an obvious restriction of the locations where they compete for their livelihoods. Moreover, new consumption patterns have sent the informal,

Fig. 9.1 Agricultural market in Caojia Alley before demolition. *Source* Photos taken by the author in 2013

undeveloped sectors of the city into a downward spiral. Noting the numbers and living demands of the rural–urban migrants in the city, Wu and Gaubatz [15] thus argued that demolition and rebuilding cannot eliminate urban poverty but will further displace the urban poor to the informal settlements at the urban fringe because of their demand for unregulated living and working spaces.

> For us, businessmen, the spaces for setting up stalls are increasingly dwindling. They were always cleansing them from the city. After they demolished them, we would find an alternative. It is increasingly hard to find a proper place. You see, the Wusi Factory has been demolished; so has Fangzheng Road. Now it is Caojia Alley. It is said after the reconstruction of the Caojia Alley, it will come here, to the Worker's Village. Everywhere is under demolition. I think they should give us some living spaces. (R38, displaced resident)

By suggesting that gentrification engenders so-called neoliberal urbanism, Smith [8] reminded us that the marginalized can be disadvantaged from the recast role of the neoliberal state in new urbanism, which no longer sustains the social reproduction of the labor force within the city but, with a global horizon, has great influence in absorbing productive investments and labor that are either present or absent from the city. Geographically, this phenomenon could result in contradictions between urban redevelopment in the metropolitan center forcing up land prices and a population marginalized from the globalized production system that can reside only in the urban periphery while their wages are earned in the city proper. The disadvantages caused by the changing role of the state from sustaining reproduction to fostering production may provoke social outrage, which may then prompt the government to strengthen territorial governance or heighten state authoritarianism in the remaining old neighborhoods of the inner city [8, 10].

In the Workers' Village (i.e., the displacee neighborhood), a new market management committee supported by the government agency at the subdistrict level (Street Office) is responsible for reconciling disputes among vendors in land-grabbing and managing the environmental health of the market. Most frequently, the conflicts involve competition for vending sites in the neighborhood and experiences regarding arbitrary charges and eviction by city managers. The city managers may require a vendor to buy newspapers from them in exchange for a stall, or they may charge for unspecified service fees. Otherwise, they may avoid setting up stalls on days when the district or municipal officials conduct inspections. These unofficial actions have largely threatened the stability of the migrant workers' income. The social control by the city managers further strengthens migrant passivity, as they have largely tended to avoid participating in urban affairs because of a fear of being evicted by the street managers, even from the remaining old neighborhoods. The provoked outrage, however, might have also induced them to talk with me, as they viewed me as someone who could perhaps speak for them and inform the public of their unfair treatment. However, every time they were asked to express themselves publicly, they held back.

With all these concerns, from the perspective of low-income migrants, the defining characteristics of gentrification are clearly perceived and portrayed. According to low-income migrants, urban redevelopment implies the upgrading of the geography of consumption and residency and consequently soaring consumption prices and rent. In addition, it signifies the emergence of land-use formalization to accommodate high-income consumers, the obsolescence of traditional consumption patterns, and the existence of constant displacement pressure on the lowest level workers in the city.

R48: Now every marketplace is formalized. Rent on the formal market is too high. It is okay if the location is good. However, if not, you can hardly balance your rental cost. For example, if there were five or more people like us who sold fish in the formal market, it was too competitive. In particular, the formal market is often only for one community, and the consumers are too limited. The informal market is much better because it is located on the open street and serves the communities all around. Currently, people living in the high-rise apartments are too lazy to go to this market. They like the shopping mall. They hope to buy cooked food and clean vegetables in the supermarket. It does not matter that the price is one yuan higher than the fresh vegetables from the farmer.

Researcher: So what do you think is the essential meaning of demolition and removal for you?

R48: Demolition means an increase in prices—everything—the house, stalls, fish, clothes, and vegetables. Everything goes up. (R48, displaced resident)

9.3 Homeownership and Urban Identity

Enduring residential displacement, increasing living costs and income instability in the city, there, the low-income migrants face the risk of social stagnation. Conceivably, housing and education are considered by migrants to be the two most important investments for them to be able to settle down in the city. Nevertheless, the residential mobility of migrant tenants with few tenure changes and a high spatial concentration reflects the relative stability of their socioeconomic status in the city. Because the housing system was public during the socialist period, inner cities in China have contained a large proportion of affordable rental housing. Urban redevelopment is associated with a sharp decrease in the inner-city affordable rental market, which contributes to the increase in housing and rental prices. Thus, although the migrants were subject to high mobility with respect to their residences and businesses, very few households purchased private apartments in the city (see also [5]). Compared with poorly educated workers, young migrants gaining specific skills or earning a degree from a secondary vocational school were more likely to be able to afford a housing mortgage in the city. With the mortgage, they have purchased secondhand apartments adjacent to their previous rental housing or workplaces.

Despite their plight, all of the informants, counterintuitively, indicated that they did not expect to obtain an urban *hukou* in Chengdu. A frequently mentioned motive for maintaining the rural *hukou* was to acquire the property from their rural land and housing. However, the more underlying reason appeared to be that these tenants considered a decoupling between citizenship and their city housing opportunities. This finding is confusing, as an urban identity may be expected to positively increase city housing opportunities given that public policies have directly associated the housing careers of individuals and households with the citizenship entitlement. A paradox emerged, which the residents explained by saying, "If I can afford housing in the city, who cares about an urban *hukou*? If the government offers me an urban *hukou*, but I still cannot afford housing, what's the use of the *hukou*?" (R49). From the migrant perspective, housing availability and *hukou* status are not interconnected and do not mutually strengthen each other. Rather, housing status, specifically home-ownership, is more likely to determine *hukou* status. *Hukou* status loses significance when private housing is unaffordable. This perception is also reflected in a passive attitude toward state assistance for low-income housing and, generally, extremely low faith in the individual's relationship with the state. The migrants argued that applying for newly built subsidized or public housing is an "unrealistic" choice compared with housing by households. As one migrant stated, "farmers are self-sufficient; how long do you think we can wait for the state to care for people like us?" (R54).

Thus, the shortfalls in the low-income housing system have not only reduced the meaning of an urban *hukou* for migrants but have also intensified the perceived significance of owning market housing in giving migrants access to the city. Wang and Fan [11] showed that, as the *hukou* continues to be a fundamental institutional barrier blocking migrants from attaining life chances equal to those of urban residents, homeownership of commodity housing can particularly strengthen the sense of identity integration in the city. Moreover, according to policies presently in force, investing in commercial apartments acts as a migrant passport to an urban *hukou* [16]. Immediately after being displaced from the CJA neighborhood, two of the 20 displacees (aged approximately 35 years old) purchased secondhand apartments in the city. One is a barber who trained in a barbershop for nearly ten years; the other is self-employed in a catering business that been in Chengdu for 16 years. Both made down payments in part by selling housing purchased much earlier in urban areas close to their rural hometowns.[2] Both residents, however, retained at least one of the family members' rural *hukou* while registering their children as local permanent residents. The double identities offer migrant households more resilience in the face of personal and family development. Homeownership, in this sense, not only eliminates the institutional barrier to urban citizenship but also enhances the life chances for

---

[2] Their parents sponsored the former private apartments of the two young adults.

migrants. In our conversations, the two informants exhibited pride in their financial success and confidence that the next generation would achieve a better life than they had, owing to the educational opportunities they have created for their children.

> I do not want to convert my *hukou* (from rural to urban registration); perhaps my husband will. Now, a rural *hukou* is much more valuable because once you are a citizen, you by no means can return to the rural land. My family has built a house on our land. We hope city development will soon expand to our village. Additionally, my mother may one day return home to live on the land, although it is small. Because we are a rural population, we could raise two children. However, our children will be part of the urban population, following their father. They must have equal educational opportunities as children in the city. They must not be like us, who suffered from poor educational attainment. (R31, displaced resident)

Nevertheless, most of the other migrants from the parental generation have expressed their disillusionment with being successful in the city. By success, they mean earning sufficient money to settle down in a private house in the city. They believe that they will ultimately return to their hometowns. The settlement intention of migrant workers as permanent citizens in central cities of China, according to [3], is set back by not only an unobtainable urban *hukou* but also unwelcome experiences in both the labor market and social interactions in the city (see also [4]). Remaining in the city is no longer the dream of older generations but implies an ingrained conviction toward the educational and job opportunities for their children in the city. In addition, the meagre agricultural income and depopulation in rural areas deter the older generations from returning to their home villages [2]. When urban policies tend to increase costs or establish other types of schooling obstacles, individuals among the older generations are increasingly depressed. For instance, according to Chengdu Municipal Government policies, since 2014, the children of migrant households have been required to purchase several kinds of social security to be exempt from charges for the nine years of compulsory education ([1], No. 2). These dilemmas often force migrants to make compromised decisions about returning to the urban areas of their hometowns.

> R54: Now, you are fortunate enough if government policy allows you to do business in the city. Success? No one imagines that any more.
>
> Researcher: So which kind of work do you think can bring you success?
>
> R54: I don't know. I thought a lot, tried a lot. But I don't know. Perhaps we can leave the question to our next generation. If I cannot create more opportunities for him, I hope he can do it on his own. Do not expect to make extra money by doing business; just keep up with your daily necessities. Just during these years, it has become harder and harder. Five years ago, the situation was still somewhat hopeful. (R54, displaced resident).

For the other city residents, ironically, their singular largest source of wealth is the hope of the awaited compensation for the requisition of their land and housing in their rural hometown. They may have improved their rural residence by virtue of its location now being in an urban area or on the rural fringes or by adding structures to their former houses, expecting excellent compensation on demolition day. For instance, a vendor living in an old neighborhood has renovated two buildings, waiting for property acquisition due to the construction of the second new international airport in Chengdu. The land on which the farmers previously relied for farming has now become a capital asset and the only asset sustaining the families. However, this expectation may act as another misconception that will finally awaken farmers. One of the street vendors in the Workers' Village has for five years been temporarily residing in an apartment on the edge of the city after his rural house was demolished. The rental fees were paid by government subsidies as part of the compensation. The vendor has pinned his family development hopes for the near future on the outstanding government compensation, which includes five apartments. The man planned to keep two of the five apartments for his parents and his own family. By selling another two apartments, he thought he could rent a space in a formal marketplace to launch a sustainable self-employed business. Optimistically, he also intended to sell his apartment in exchange for a smaller one with a better location for the schooling of his expected baby.

9.4   Conclusion

Part Three discusses the unevenness in the social effects of state-facilitated gentrification in Chengdu (Table 9.1). State dominance in conferring variable housing opportunities post-gentrification among social groups by housing policy-making has determined the significant disparity of impacts from urban redevelopment in China. After the post-gentrification housing and compensation schemes, the former public tenants and subsidized owners associated with socialist work units were privileged through cash compensation or relocation to receive new, self-owned apartments on- or offsite. We also reveal the social transformation of the socialist workers from welfare recipients in *danwei* compounds to proprietors in the new, gated communities. The migrant workers, on the other hand, suffered the unmitigated assault of displacement without compensation. A third group, owners of commodity or informally self-built housing, has faced a process of forced lifestyle change following the property rights violation. Normally, this housing class has engaged in difficult negotiations with the state over appropriate compensation [7].

The social change of the three social groups responses to the sociological doctrine that stands for the persistence of administrative power in the market society. The socialist workers have taken advantage of their previously political status to access higher and better housing opportunities. As a result, a new housing class is in formation, distancing itself from impoverished private tenants. *Hukou* system, on the other hand, continues to play in regulating the redistribution of spatial resources.

The policy-making of property expropriation, compensation and resettlement essentially reflects the complex bundle of state–society relations. First, for socialist workers and some relocatees who pursued the maximization of economic benefits, the state–society relation is a positive-sum relationship, which, however, caused externally negative effects, that is, an increase in urban density. Based on the solid political foundation of this group and the institutional environment formed in the planned economy, this group has the most easily cooperated with the government and

Table 9.1 The experiences of three low-income groups in gentrification

	Identities	Compensation	Perceptions	Results
Socialist workers	• Citizens • Public tenants/subsidized owners	• Profit concession (homeownership promotion)	• Personal transformation from proletariats to propertied social stratum	• Aiding consensus-building • Relocation to high-rise, gated communities • Changes in consumption behaviors • Consumption and displacement pressure
Activists	• Citizens/migrants • Homeowners of commodity housing, historic dwellings, and informal buildings	• Property exchange or monetary compensation	• Property rights violation • Forced lifestyle change • Emotional damage	• Activism • Relocation to high-rise, gated communities • Prolonged confrontation with the government
Migrant tenants	• Rural–urban migrants • Private tenants	• Eviction without compensation	• Formalization • Spatial commodification	• Inactive • Move individually to other migrant neighborhoods • Double marginalization by the *hukou* system and the increasing consumer society

this is the group toward which the government policy has been inclined. Finally, the government and socialist workers formed the consensus-building for gentrification based on benefit-sharing.

For radical resisters, the local government has adopted a type of cultural hegemony. An important reason for easing the social resistance of the latter two groups is the existence of the former, that is, the pragmatic coalition of the state and socialist workers. This kind of pragmatic coalition is manifest in a series of soft management means, such as persuasion, emotional guidance, benefit induction, etc. This kind of cooperation between government and society is different from state corporatism: first, its social object is vulnerable rather than elite groups; second, it is not based on fixed organizations and institutional frameworks but on a loose, temporary alliance to achieve collective goals.

Finally, considering urban immigrants, the local government continues to act through a simple administrative authority whose operation benefits from the long-standing dichotomy between urban and rural societies and the resulting internalization of the rural household registration population in recognition and behavior.

The complex state–society relationship causes specific social results of gentrification in Chinese cities: State-facilitated gentrification in Chinese cities is accompanied and mutually reinforced by state-led social transformation of part of the lower-income groups in the market society. In this case study of Chengdu during the 2000s, onsite relocated socialist workers were included among the gentrifiers, while migrant workers and owners of commodity and informal housing suffered negative gentrification effects in the form of either direct housing and economic losses or emotional damage due to forced lifestyle changes.

References

1. Bureau of Education of Chengdu (2014) Guanyu zuohao jincheng wugong renyuan suiqian zilu jieshou yiwujiaoyu gongzuo de yijian (Opinions on works for the compulsory education for the children of migrant workers), No. 2. Retrieved 13 Aug 2015, from: http://www.cdzk.com/upload/Content_Publish/sqzv/tz.pdf
2. Chan KW (2010) The household registration system and migrant labor in China: notes on a debate. Popul Dev Rev 36(2):357–364
3. Fan CC (2011) Settlement intention and split households: findings from a migrant survey in Beijing, China. China Rev 11(2):11–42
4. Fan CC, Sun M, Zheng S (2011) Migration and split households: a comparison of sole, couple, and family migrants in Beijing, China. Environ Plan A 43(9):2164–2185
5. Li S, Zhu Y (2014) Residential mobility within Guangzhou city, China, 1990–2010: local residents versus migrants. Eurasian Geogr Econ 55(4):313–332
6. Pun N (2006) Shiyuzhe de hushing—Zhongguo dagongmei koushu (The voice of the aphasic: the oral statement of Chinese female migrant workers). Beijing Sanlian Publishing House
7. Shin HB (2013) The right to the city and critical reflections on China's property rights activism. Antipode 45(5):1167–1189
8. Smith N (2002) New globalism, new urbanism: gentrification as global urban strategy. Antipode 34(3):427–450
9. Solinger D (1999) Contesting citizenship in urban China. University of California Press, Berkeley
10. Swyngedouw E (1997) Neither global nor local: "glocalisation" and the politics of scale. In: Cox K (ed) Spaces of globalisation: reasserting the power of the local. Guilford Taylor P, New York, pp 137–166 (1995) World cities and territorial.
11. Wang W, Fan CC (2012) Migrant workers' integration in urban China: experiences in employment, social adaptation, and self-identity. Eurasian Geogr Econ 53(6):731–749
12. Wu W (2002) Migrant housing in urban China: choices and constraints. Urban Aff Rev 38(1):90–119
13. Wu W (2006) Migrant intra-urban residential mobility in urban China. Hous Stud 21(5):745–765
14. Wu F (2012) Neighborhood attachment, social participation, and willingness to stay in China's low-income communities. Urban Aff Rev 48(4):547–570
15. Wu W, Gaubatz P (2013) The Chinese cities. Routledge, New York
16. Zhang L, Wang G (2010) Urban citizenship of rural migrants in reform-era China. Citizsh Stud 14(2):145–166

Chapter 10
State-Facilitated Gentrification as a Transformative Force of Development

After the 1998 Asian crisis and certainly since the end of 2001, when China joined the World Trade Organization, many large Chinese cities have competitively unleashed strategies to achieve a fast track of globalization. These post-socialist cities, however, are also confronting the sizeable expansion of their low-skilled service employment base. Labor participation is derived from an annually increasing number of migrants leaving their rural hometowns and seeking opportunity in the urban terrain. Meanwhile, these cities have actively relocated their manufacturing base from central areas to new locations and have attempted to convert their downtowns into centers of global culture and high-end services. However, the degree of postindustrial advancement has been moderate. In the meantime, a new social stratum taking advantage of market reform has earned considerable personal wealth. Small businessmen, especially those in township enterprises established in the rural homeland, have gained greater policy support for development. Meanwhile, many pre-reform elites in urban areas have experienced job transfer and have readapted their personal skills to maintain their elite status after the turbulence generated by the labor market reform. Additionally, young professionals in rising sectors have experienced greater competition in the new market. These new, rich cohorts in China thus feature heterogeneous cultural capital, lifestyles and politics stemming from the different social characteristics cultivated under different economic systems. The transformation from a production to a consumption economy has accompanied social policies by the post-Mao state that have shifted from developing a mass society to developing a middle-class society, which has been termed a "moderate prosperity society." Under this political ideology of socioeconomic development, individual consumption is commonly mobilized among China's new rich.

The above review portrays a larger society that differs from that of the postindustrial society in the West. However, the trends of economic marketization and the emerging neoliberal ideas of the reformist state clearly coincide with many global metropolises. Derived from these contexts, the structural tendency of sociospatial upgrading in the central city of Chengdu is not necessarily an ascendant urban process. Gentrification is a unique urban trend that is normally juxtaposed against

© Springer Nature Singapore Pte Ltd. 2022
Q. Yang, *Gentrification in Chinese Cities*, Urban Sustainability,
https://doi.org/10.1007/978-981-19-2286-2_10

changes in other localities adapting to the expansion of low-skilled service workers in the central city. And it has been selectively fostered in fewer localities in the central city. The formation of gentrification should rely on leveraging innovative urban strategies to turn the tide of urban transformation in the inner city.

This study documents three sets of state interventions that have guided gentrification and residential relocation. The first two were innovative urban strategies that initiated gentrification, and the last was a countermeasure necessary for a social cure post-gentrification. They are related to the institutional changes that were implemented for the creation of new cultural urbanism, spatial commodification, and the redistribution of spatial resources among lower-income social groups.

10.1 Creative Destruction of Consumption Spaces

Concerning the first strategy, in the newly developed city, where globalization in Chengdu is an imaginary more than an existing phenomenon, the inner-city terrain is at the cutting edge, allowing the politico-economic elites to *iteratively* (re-)invent the landscape and space to (re)market the city to the world. Initially, the landscape change in Chengdu reflected a trend of aestheticization in which the symbols of the European and South Asian cityscape offered an imaginary of cultural globalization and were delivered as promising a better life for the new rich. In this wave of city branding and spatial production, a newly built mode was dominated the style of, in particular, living space reconstruction. Subsequently, given the urban image desired by political-economic actors, a new trend appeared in which extremely modernist and postmodernist landscapes suggested a higher level of development in modernization and globalization. These modernist and postmodernist landscapes showed a Chinese city that tended to be culturally and aesthetically more integrated and inclusive in a time of cosmopolitanism. The rennovated inner-city spaces were depicted by developers as representing the elite lifestyles, which have ultimately catered to transnational investors, business people, and high-income consumers in the city.

Cultural place-making was reinforced by a process of consumer class-making, paralleling the sociocultural transformation from a decommodified to a consumer society. Ideologically, the place-making of gentrified neighborhoods is connected to cultural modernization and globalization but also integral with a local identity that caters to high-income consumers. The Chengdu gentrifiers have tended to pursue cultural prestige as symbolized by the modernist spectacle and the historical places in the city center. As a result, the new urbanism in the inner city supports gentrifiers as they accumulate a high level of sociocultural capital. High cultural and economic capital have thus become the main discourse for inner-city gentrifiers. Living in gentrified places, gentrifiers could then cultivate comparable consumption behaviors and collective recognition.

Indeed, the fervent desire for global city-making may wane, resulting in more rationality in the use of global and local symbols in spatial production and more prudence in old-community cleansing. However, the creative production of new consumption spaces, either residential or retail, which drive the formation of new consumer cultures, is sustained as a basic urban development strategy.

10.2 Excess Spatial Production

In combination with the changing cultural code of place-making, rule-making for land reuse promotes a mode of speculative surplus accumulation by overproducing consumption spaces, which necessitates that spatial consumption be encouraged. A common account focuses on the energetic role taken by the state and land developers to shift their form of accumulation from industrial development to the built environment, a change that has been authorized by the public land system in China. This study strengthens this argument by providing evidence of the speculative practices of land capitalization and the boosting of the real estate market. In China, land marketization via urban redevelopment implies not only the appropriation of the rent gap but also the fulfilment of state legitimacy in a time of market transition. The conflated economic and sociopolitical objectives encouraged the state to adopt land-reuse rules that ultimately intensified spatial commodification and commercialization, such as the increase of plot ratio, the commercial land-use proportion, and high-end property or retail development. The result was the state-led expansion of the rent gap. This intensified spatial commodification and commercialization assumed a standing army of consumers who, once production was completed, would consume. The initiation of urban redevelopment was thus consistently accompanied government policies acting as stimuli to housing consumption. The excess production of housing and retail space finally resulted in dependence on a consumer-driven economy.

Similarly, the era of property-led gentrification is passing. China's property construction and consumption booms are fading. Instead, local governments are shifting toward encouraging daily consumption. The new gentrification trend will feature more excess production of retail spaces to catalyze new consumption types.

10.3 Unequal Spatial Resource Redistribution

Following the inner-city settlement by the middle class, the residential relocation of the lower-income residents has been subject to oversight from government policies integral to the property expropriation laws. State intervention in the residential relocation of affected residents reflects the political ideology of maintaining social stability and the Chinese state tactics of social governance (see also [4]). In practical terms, the central and local governments have traded economic growth for social solidarity by deploying various compensation schemes to address the housing losses of former occupants through urban redevelopment. The implementation of different compensation methods has diversified the outcomes of social groups in the relocation process and has added layers of meaning to urban society.

First, among these methods, most noticeable was a decision over public–private tenure conversion for low-income socialist workers. This decision enabled home-ownership access for public tenants, which imparted social legitimacy to the state and corporate actors managing the urban redevelopment and residential relocation. As a result, the group of residents could become gentrification stakeholders and even, gentrifiers. Nevertheless, the retreat of a welfare state and the status change to proprietors might unleash both market success and failure for some new homeowners, anticipating new inequalities in the market society.

Private owners, who reside in either historic places or informal dwellings, were offered a so-called equivalent exchange for their old property with new property or cash. With a strong consciousness of property rights and personal lifestyle, these property owners became the most active resistors of displacement and property invasion and the defenders of cultural diversity in the city. However, the partici-pation of public tenants in social governance greatly frustrated the power of private owners to resist. Despite a sense of displacement, these homeowners ultimately became a new type of urbane citizen by entering gated communities and high-rise apartments. Finally, a still-powerful *hukou* system continued to disempower marginalized migrant groups, preventing them from challenging the formal decision-making system. Private tenants, who were almost always rural–urban migrants, were unsurprisingly denied any legal right to compensation and evicted from their neighborhoods.

10.4 Synthesis

Synthesizing the above three processes, this study suggests that state-facilitated gentrification in Chengdu should be read as a transformative force of development led by the state and capitalists and championed by middle-class consumers. The objec-tives of state-facilitated gentrification might originate from industrial upgrading, land finance, talent attraction or simply the expansion of domestic demand, but they ultimately lead to the cycles of transformative effects on economic and social devel-opment through spatial production. This gentrification mode catalyzes new spaces and collective cultures, which then necessitate the stimulation of new consumption behaviors and the formation of new consumer classes, at the expense of the spatial demands for the even larger number of low-income residents. However, in the context of China's unique state–society relations, some low-income groups may also ride the wave of social transformation. This type of gentrification has a much wider connota-tion than the issue of place-based class conflicts or the exploitation of land rent gap by capitalists. It integrates into not the essence of uneven geographical development in a capitalist society, but China's unique model of urbanization and development [2], which is often state-driven, innovative and even involuted so as to sustain continuous growth.

Based on this epistemology, the in-depth study in Chengdu points to four territo-rial gentrification characteristics that could be shared by other large Chinese cities.

First, while state-led urban strategies provided the initial impetus for gentrification in Chengdu, the consumer revolution has awakened personal consumption as a force sustaining its development. Politico-economic and sociocultural forces are both indispensable in advancing gentrification in China. Second, a consumer class, with common consumption preference, replacing a socioeconomic class, has nurtured class-related urban change in certain inner-city places during the 2000s. Moreover, as new consumption spaces are produced by the local state and capitalists and as consumer demands increase, middle-class consumers will continue to divide in their cultural identification and form new cultural forces supporting the new geography of gentrification. Consumer class-making has been dynamically interwoven with place-making in the gentrification process. Gentrification thus facilitates place-based social change in the post-socialist city rather than occurring as an outcome of postindustrial societal transformation. Third, to smooth gentrification, the state has also tended to guide the working-class transformation. Fourth, and as a result, state-facilitated gentrification could cause uneven social effects by launching unequal institutional arrangements for housing lower-income social groups.

This book argues that the interactions of state and society, rather than state or society in domination [3, 8, 12], have determined these gentrification characteristics. Such state–society interactions are significant complicating factors for gentrification politics in Chinese contexts, rendering it a neutral rather than simply a critical process. On the one hand, state-facilitated gentrification in Chengdu has united the gentrifiers to fit functionally with the new market regime; together, they model middle-class urbanism. During the process of socio-spatial upgrading, state action in spatial production directs the change in the consumer cultures and behaviors of middle-class newcomers to the inner city to be compatible with state modernization strategies, economic restructuring and real estate boosterism. It shapes a subjectively constructed path of consumer citizens who are attendant to elite-oriented place-making. This gentrification process contributes to the reproduction of *conservative politics* within the new rich in the post-socialist societal transition, which is compared with the politics held by progressive urban pioneers in the early gentrification stage in postindustrial cities.

On the other hand, the local state and some residents previously subject to the *danwei* system have formed a temporary and pragmatic coalition to mitigate social conflicts, which could be deemed a *quasi-state corporatism*. Through this coalition, a part of the local residents have become de facto gentrifiers, sharing land increments with capitalists and other newcomers. The real displacees are private owners and migrant tenants, who suffered from the *cultural hegemonic* and *authoritative politics* of the state. The hegemonic politics are manifested in the demolition of the prerevolutionary and socialist dwellings and informal housing and the normalization of housing consumer cultures with a spatial prototype of respectable and governable gated communities. The authoritative politics ultimately reinforced the institutionalized marginalization of the migrant tenants.

10.5 Limitations

Finally, the spatiotemporality of this study should be discussed. Several key common-alities among large Chinese cities are treated with caution throughout the study. First, the stages of socioeconomic restructuring in contrast with the post-industrial societies and the background of consumer revolution in the transitional society are compa-rable among the four large cities discussed in this book. Then, roughly from 1990 to the early 2010s, the globalization strategies were prevalent in national and regional central cities in China, which sparked off large-scale urban renewal programs. The local governments' cultural strategies reflected in these urban renewal programs in these cities is analogous to each other. Also, in the renewal projects, similar means stimulated land reuse, such as land tax relief and the plot-ratio adjustment. Last, the national and subnational governments had a high degree of unity in the changing policies of property expropriation and compensation.

However, different cities do have different degrees of changes in their social and economic structures and in the development of a consumer society. For example, because Chengdu is the capital city of Sichuan Province, the growth rate of intraprovincial migrants to Chengdu is salient. This rate represents a demographic trend resisting gentrification. Also, the changing cultural strategies of urban renewal might appear differently in time and in extent. Moreover, distinctions exist between the powers and responsibilities of political and economic forces. Compared with Shanghai and Beijing, known as "world cities" in China, the Chengdu case could have shown a relatively significant role for the local state in leading the way in gentrification, as well the other types of urban projects. The transnational capitalist influence has likely countered state power in the gentrification process in Shanghai and Beijing. The large number of real estates remaining in the Chengdu inner city, which are owned by state-owned *danwei*s, could be another factor boosting the role of local state actors.

Finally, the local groups and the types of housing targeted by gentrification could vary by city and even by case. Most obviously, the large number of rural housing areas in Shenzhen and Guangzhou changed the initial social and physical conditions of gentrification. In the 2000s, rural property owners became the most aggressive forces of resistance in these cities (see [7]), similar to the private property owners in this book. However, as the revolt progressed, the city government began to reach a consensus with some of these rural property owners. The government formulated policies such as onsite resettlement and high-priced compensation to obtain the rural property owners' consent to renewal while relaxing the plot ratio control to increase real estate development. This change in the institutional design of the residential relocation of rural homeowners is comparable with the government's approach to socialist workers.

The study thus deems the backgrounds of a transitional society and national and local policies and institutions involved in gentrification as homologous among different large Chinese cities. The homology underpins the book's explanation for gentrification particularly during the 2000s. However, the city-level differences in the degree of development, politico-economic conditions and project-based social and physical conditions could result in the variations of the degree to which gentrification is advanced, the social participation in it, and both the types of cultural place-making and the degree of displacement it causes.

Concerning with temporality, the study concentrated on the 2000s, and, in individual cases (e.g., CJA), the early 2010s. Urban construction during these times showed the rapid expansion of the city and the predominance of the newly built mode of urban renewal. Urban modernization construction remained in its early stage. Most of the new buildings were inspired by the architectural aesthetics of developed countries [1, 6, 9]. In 2011, China's urbanization rate reached 50%, after which the central work of urban planning began to shift to the management of the urban built-up areas instead of land-use expansion. At the same time, the deepening of reform of economic system also resulted in a shift in urban development strategies from extensive growth to quality improvement (State [10]). In this context, the era of large-scale demolition and reconstruction waned in major cities. The latest urban renewal policies in the central and local governments underlined the environmental renovation, functional replacement, facility supplement and community governance of the old urban areas ([5, 13], No. 243). At the same time, high-tech industries have advanced in large city economies. Major cities have radially launched talent-introduction strategies.

In this context, gentrification began to transform from large-scale, newly built residential gentrification to small-scale, superhigh-rise residential and commercial gentrification. Spontaneous commercial gentrification has also developed rapidly. The new forms of gentrification hopefully encourage more studies on its new characteristics and effects. Notwithstanding, this round of urban renewal features an even greater level of national state encouragement and policy guidance, which covers a much larger scale in the old cities than simply dilapidated areas (State [11], No. 23). Gentrification will thus continue to deliver the state imaginary of a new type of urbanization and cityscape. However, the state imaginary will be more related to the combination of global and local cultures and the balance between development and historical preservation. Also, the land-reuse policies in urban renewal are expected to depart from previously land expropriation system and leads to quality improvement and partially functional adjustment of previous land-use. However, the creation of new consumption space still constitutes a pervasive purpose of different projects in this new round of city regeneration.

In this process of urban regeneration, new consumption spaces will be produced and embraced by the newly forming consumer groups. These spaces feature emerging retail spaces in either the postmodernist historical streets or the glass-trimmed shops in the old residences and large flat floors of inner-city skyscrapers. The displacement process, on the other hand, is expected to occur in a more exclusionary way than does wholesale relocation. The changes and invariances thus endorse the analytical frame of this book and are useful for explaining temporal gentrification. However, the sociocultural evolution of China's consumers; the extremely elastic policy-making by the government intervening in the housing market; and, perhaps most influentially, the continual socioeconomic restructuring and penetration of transnational forces (of both production and consumption) are all expected to provide new empirical materials that will continue to intensify and transform our understanding of the global–local and spatiotemporal relations implied within gentrification.

References

1. Farrell K, Westlund H (2018) China's rapid urban ascent: an examination into the components of urban growth. Asian Geogr 35(1):85–106
2. Hamnett C (2020) Is Chinese urbanisation unique? Urban Stud 57(3):690–700
3. He S (2007) State-sponsored gentrification under market transition: the case of Shanghai. Urban Aff Rev 43(2):171–198
4. Lin GCS (2015) The redevelopment of China's construction land: practising land property rights in cities through renewals. China Q 224:865–887
5. Ministry of Housing and Urban-Rural Development (2019, Apr) Guanyu zuohao 2019nian laojiuxiaoqu gaizaogongzuo de tongzhi (Notice on the Renovation of Old Residential Areas in 2019), No.243. Retrieved 05 Jan 2022, from. http://www.scio.gov.cn/ztk/38650/40922/index. htm
6. Schneider A, Chang C, Paulsen K (2015) The changing spatial form of cities in Western China. Landsc Urban Plan 135:40–61
7. Shin HB (2013) The right to the city and critical reflections on China's property rights activism. Antipode 45(5):1167–1189
8. Shin HB, Lees L, López-Morales E (2016) Introduction: locating gentrification in the global east. Urban Stud 53(3):455–470
9. Sorichetta A, Nghiem SV, Masetti M, Linard C, Richter A (2020) Transformative urban changes of Beijing in the decade of the 2000s. Remote Sens 12(4):652
10. State Council (2014, Mar) Guojia xinxing chengzhenhua guihua (2014–2020) (National Plan on New Urbanization 2014–2020). Retrieved 05 Jan 2022, from: http://dzb.scdaily.cn/2014/03/17/20140317620143994732.htm
11. State Council (2020, July) Guangyu quanmian tuijin chengzhen laojiuxiaoqu gaizaogongzuo de zhidao yijian (Guiding Opinions on Promoting the Renovation of Old Urban Residential Areas in an All-round Way), No. 23. Retrieved 05 Jan 2022, from: http://www.gov.cn/zhengce/content/2020-07/20/content_5528320.htm
12. Wu F (2016) State dominance in urban redevelopment: beyond gentrification in urban China. Urban Aff Rev 52(5):631–658
13. Yang JQ, Chen Y (2020) 1949–2019 nian zhongguo chengshi gengxin de fazhan yu huigu (Review of the development of urban renewal in China from 1949 to 2019). Urban Plann 44(02):9–19 + 31

Correction to: Gentrification in Chinese Cities

Correction to:
Q. Yang, *Gentrification in Chinese Cities*, Urban
Sustainability, https://doi.org/10.1007/978-981-19-2286-2

In the original version of the Book Frontmatter, a wrong word was added in the Funding Information text which has now been updated. The Book Frontmatter has been updated with the change.

The updated version of the book can be found at
https://doi.org/10.1007/978-981-19-2286-2

Appendix A
An Overview of Fieldwork Locations

Informant group	Field site	Number of informants
Onsite relocated residents	Caojia Alley	4
	Jinniu Wanda	16
Offsite relocated residents	Qingxi Yazhu	1
	Quanshui Renjia	10
	Donghong Guangxia	4
	Jinxiu Dongfang	5
Activists	Caojia Alley	1
	Jinxin Jiayuan	4
	Wide and Narrow Alley	5
Displaced migrant tenants	Workers' Village	18
	Balizhuang	2
Residential gentrifiers	Times Riverside	4
	Chengdunese Paradise	4
	Waterfront	4
	Jintianfu	8

© Springer Nature Singapore Pte Ltd. 2022
Q. Yang, *Gentrification in Chinese Cities*, Urban Sustainability,
https://doi.org/10.1007/978-981-19-2286-2

Appendix B
An Overview of Informants

Code	Informant group	Affiliation	Field site	Date
First stage				
O1	Urban planner	Institute of Planning and Design	Chengdu	2014/12/19
O2	Manager	Task Force of Urban Redevelopment	Jinniu District	2014/12/23
O3	Committee member	Community Residential Committee	Caojia Alley	2014/12/25
R4	On-site relocated resident	Resettlement community	Caojia Alley	2014/12/25
R5	On-site relocated resident	Resettlement community	Caojia Alley	2014/12/26
R6	Off-site relocated resident	Resettlement community	Quanshui Renjia	2014/12/26
R7	Off-site relocated resident	Resettlement community	Quanshui Renjia	2014/12/26
R8	Off-site relocated resident	Resettlement community	Quanshui Renjia	2014/12/27
R9	Off-site relocated resident	Resettlement community	Quanshui Renjia	2014/12/27
R10	Off-site relocated resident	Resettlement community	Quanshui Renjia	2014/12/28
R11	Off-site relocated resident	Resettlement community	Donghong Guangxia	2014/12/29
R12	Off-site relocated resident	Resettlement community	Donghong Guangxia	2014/12/31
R13	Off-site relocated resident	Resettlement community	Donghong Guangxia	2015/01/05
R14	Off-site relocated resident	Resettlement community	Donghong Guangxia	2015/01/05

(continued)

© Springer Nature Singapore Pte Ltd. 2022
Q. Yang, *Gentrification in Chinese Cities*, Urban Sustainability,
https://doi.org/10.1007/978-981-19-2286-2

(continued)

Code	Informant group	Affiliation	Field site	Date
O15	Committee member	Community Residential Committee	Jinniu Wanda	2015/01/06
R16	Activist	Caojia Alley	Caojia Alley	2015/01/08
R17	On-site relocated resident	Resettlement community	Jinniu Wanda	2015/01/08
R18	On-site relocated resident	Resettlement community	Jinniu Wanda	2015/01/09
R19	On-site relocated resident	Resettlement community	Jinniu Wanda	2015/01/10
O20	Urban planner	Institute of Planning and Design	Chengdu	2015/01/10
R21	On-site relocated resident	Resettlement community	Jinniu Wanda	2015/01/12
O22	Manager	Urban–Rural Housing Department	Jinniu District	2015/01/13
O23	Manager	Urban–Rural Housing Department	Jinniu District	2015/01/13
O24	Manager	Financing Platform Company	Caojia Alley	2015/01/14
R25	On-site relocated resident	Resettlement community	Jinniu Wanda	2015/01/15
R26	Displaced migrant tenant	Displacees' community	Workers' Village	2015/01/22
R27	Displaced migrant tenant	Displacees' community	Workers' Village	2015/01/23
R28	Displaced migrant tenant	Displacees' community	Workers' Village	2015/01/23
R29	Displaced migrant tenant	Displacees' community	Workers' Village	2015/01/23
R30	Displaced migrant tenant	Displacees' community	Workers' Village	2015/01/24
R31	Displaced migrant tenant	Displacees' community	Workers' Village	2015/01/25
R32	Displaced migrant tenant	Displacees' community	Workers' Village	2015/01/25
O33	Manager	Task Force of Urban Redevelopment	Chenghua District	2015/01/27
O34	Manager	Danwei (property owner)	Caojia Alley	2015/01/27
O35	Manager	Danwei (property owner)	Caojia Alley	2015/01/28

(continued)

(continued)

Code	Informant group	Affiliation	Field site	Date
R36	Displaced migrant tenant	Displacees' community	Workers' Village	2015/02/02
R37	Displaced migrant tenant	Displacees' community	Workers' Village	2015/02/02
R38	Displaced migrant tenant	Displacees' community	Workers' Village	2015/02/02
R39	On-site relocated resident	Resettlement community	Jinniu Wanda	2015/03/01
R40	On-site relocated resident	Resettlement community	Jinniu Wanda	2015/03/01
R41	On-site relocated resident	Resettlement community	Jinniu Wanda	2015/03/02
R42	On-site relocated resident	Resettlement community	Jinniu Wanda	2015/03/02
R43	On-site relocated resident	Resettlement community	Jinniu Wanda	2015/03/02
R44	On-site relocated resident	Resettlement community	Jinniu Wanda	2015/03/03
R45	On-site relocated resident	Resettlement community	Jinniu Wanda	2015/03/04
R46	On-site relocated resident	Resettlement community	Jinniu Wanda	2015/03/05
R47	Off-site relocated resident	Resettlement community	Qingxi Yazhu	2015/03/05
R48	Displaced migrant tenant	Displacees' community	Workers' Village	2015/03/06
R49	Displaced migrant tenant	Displacees' community	Workers' Village	2015/03/06
R50	Displaced migrant tenant	Displacees' community	Workers' Village	2015/03/06
R51	Displaced migrant tenant	Displacees' community	Workers' Village	2015/03/07
R52	Displaced migrant tenant	Displacees' community	Workers' Village	2015/03/07
R53	Displaced migrant tenant	Displacees' community	Workers' Village	2015/03/07
R54	Displaced migrant tenant	Displacees' community	Workers' Village	2015/03/08
R55	Displaced migrant tenant	Displacees' community	Workers' Village	2015/03/08
R56	Displaced migrant tenant	Displacees' community	Balizhuang	2015/03/09

(continued)

(continued)

Code	Informant group	Affiliation	Field site	Date
R57	Displaced migrant tenant	Displacees' community	Balizhuang	2015/03/09
O58	Architect	China Construction Southwest Design and Research Institute	Chengdu	2015/03/18
O59	Architect	China Construction Southwest Design and Research Institute	Chengdu	2015/03/20
G60	Retail gentrifier	Commercialised historic area	Wide and Narrow Alley	2015/03/23
G61	Retail gentrifier	Commercialised historic area	Wide and Narrow Alley	2015/03/23
G62	Retail gentrifier	Commercialised historic area	Wide and Narrow Alley	2015/03/24
G63	Retail gentrifier	Commercialised historic area	Wide and Narrow Alley	2015/03/24
G64	Retail gentrifier	Commercialised historic area	Wide and Narrow Alley	2015/03/25
G65	Retail gentrifier	Commercialised historic area	Wide and Narrow Alley	2015/03/25
G66	Retail gentrifier	Commercialised historic area	Wide and Narrow Alley	2015/03/26
G67	Retail gentrifier	Commercialised historic area	Wide and Narrow Alley	2015/03/26
G68	Retail gentrifier	Commercialised historic area	Tangba Street	2015/03/27
G69	Retail gentrifier	Commercialised historic area	Tangba Street	2015/03/27
O70	Expert	Southwest Transportation University	Chengdu	2015/03/29
O71	Expert	Sichuan University	Chengdu	2015/03/30
Second stage				
R72	On-site relocated resident	Resettlement community	Caojia Alley	2015/11/10
R73	On-site relocated resident	Resettlement community	Caojia Alley	2015/11/11
R74	On-site relocated resident	Resettlement community	Jinniu Wanda	2015/11/13
R75	On-site relocated resident	Resettlement community	Jinniu Wanda	2015/11/14
R76	On-site relocated resident	Resettlement community	Jinniu Wanda	2015/11/14

(continued)

(continued)

Code	Informant group	Affiliation	Field site	Date
R77	Activist	Wide and Narrow Alley	Wide and Narrow Alley	2015/11/20
R78	Activist	Wide and Narrow Alley	Wide and Narrow Alley	2015/11/23
R79	Activist	Wide and Narrow Alley	Wide and Narrow Alley	2015/11/23
R80	Activist	Wide and Narrow Alley	Wide and Narrow Alley	2015/11/25
R81	Activist	Wide and Narrow Alley	Wide and Narrow Alley	2015/11/25
R82	Off-site relocated resident	Resettlement community	Jinxiu Dongfang	2015/12/10
R83	Off-site relocated resident	Resettlement community	Jinxiu Dongfang	2015/12/10
R84	Off-site relocated resident	Resettlement community	Jinxiu Dongfang	2015/12/11
R85	Off-site relocated resident	Resettlement community	Jinxiu Dongfang	2015/12/11
R86	Off-site relocated resident	Resettlement community	Jinxiu Dongfang	2015/12/12
R87	Activist	Resettlement community	Jinxin Jiayuan	2015/12/15
R88	Off-site relocated resident	Resettlement community	Quanshui Renjia	2015/12/16
R89	Off-site relocated resident	Resettlement community	Quanshui Renjia	2015/12/16
R90	Off-site relocated resident	Resettlement community	Quanshui Renjia	2015/12/17
R91	Off-site relocated resident	Resettlement community	Quanshui Renjia	2015/12/17
R92	Off-site relocated resident	Resettlement community	Quanshui Renjia	2015/12/18
G93	Residential gentrifier	Gentrifiers' community	Chengdunese Paradise	2015/12/19
G94	Residential gentrifier	Gentrifiers' community	Chengdunese Paradise	2015/12/19
R95	Activist	Resettlement community	Jinxin Jiayuan	2015/12/20
R96	Activist	Resettlement community	Jinxin Jiayuan	2015/12/21
R97	Activist	Resettlement community	Jinxin Jiayuan	2015/12/22
G98	Residential gentrifier	Gentrifiers' community	Chengdunese Paradise	2015/12/27
G99	Residential gentrifier	Gentrifiers' community	Chengdunese Paradise	2015/12/27

(continued)

(continued)

Code	Informant group	Affiliation	Field site	Date
G100	Residential gentrifier	Gentrifiers' community	Times Riverside	2015/12/31
G101	Residential gentrifier	Gentrifiers' community	Times Riverside	2015/12/31
G102	Residential gentrifier	Gentrifiers' community	Times Riverside	2016/01/02
G103	Residential gentrifier	Gentrifiers' community	Times Riverside	2016/01/03
G104	Residential gentrifier	Gentrifiers' community	Waterfront	2016/01/08
G105	Residential gentrifier	Gentrifiers' community	Waterfront	2016/01/09
G106	Residential gentrifier	Gentrifiers' community	Waterfront	2016/01/09
G107	Residential gentrifier	Gentrifiers' community	Waterfront	2016/01/16
Third stage				
G108	Residential gentrifier	Gentrifiers' community	Jintianfu	2019/08/15
G109	Residential gentrifier	Gentrifiers' community	Jintianfu	2019/08/20
G110	Residential gentrifier	Gentrifiers' community	Jintianfu	2019/08/22
G111	Residential gentrifier	Gentrifiers' community	Jintianfu	2019/08/23
G112	Residential gentrifier	Gentrifiers' community	Jintianfu	2019/08/24
G113	Residential gentrifier	Gentrifiers' community	Jintianfu	2019/08/25
G114	Residential gentrifier	Gentrifiers' community	Jintianfu	2019/08/26
G115	Residential gentrifier	Gentrifiers' community	Jintianfu	2019/08/28

Note O means organisational member, G means gentrifier, R means relocated resident

Appendix C
Interview Guide

For Organisational Members Who Once Participated in the Renewal Projects

Project Information and Polices

1. How did you first hear of the project to redevelop this district?
2. Can you introduce the social composition and tenure types of the neighborhood before redevelopment?
3. What are the national and local policies that have influenced your work in the project?

Project Participation

4. Can you introduce your institution? What are your main jobs in the project?
5. Can you describe your interactions with the other parties participating in the project (e.g., *danwei* managers, officials, developers, planners and resident representatives)? Which kind of difficulties did you meet in the interactions?
6. Before the project was formally established, did you make any attempts to inform the residents or seek their opinion and advice on the project? What were the main concerns expressed by varying community members?
7. How did you mediate the contentious issues that were brought forth by residents? Can you give me an example?
8. What are the main questions you were asked in the plan making and implementation of residents' removal and compensation?
9. What are the main concerns about the spatial planning for the new neighborhood in this area?

© Springer Nature Singapore Pte Ltd. 2022
Q. Yang, *Gentrification in Chinese Cities*, Urban Sustainability,
https://doi.org/10.1007/978-981-19-2286-2

Difficulties and Opinions

10. What do you think the major challenge of the project is? How have these issues been tackled?
11. How do you think this project will impact the current residents in particular, and the city in general?

For Residential Gentrifiers

Occupation and Income

1. Can you tell me your occupation and educational background? How about your family members?
2. How does your family income compare to others in this city—richer, poorer, the same? Can you tell me generally your family's income band every year?
3. Did you experience an important job transfer? Can you tell me the reason?

Experience in Residential Mobility

4. Before living here, where have you lived? What are the tenure types of your former dwellings?
5. Can you tell me the main reasons for moving each time?
6. Do you have other housing properties now?

Housing Choice of the Current Apartment

7. How long have you been living in the current neighborhood?
8. Do you live together with your children/parents? If not, where do they live now?
9. Do you and your families enjoy living in this neighborhood? Why?
10. Did you ever consider living in other places in Chengdu?
11. Can you tell me more about your neighbours? Do you often interact with the neighbours?
12. Have you participated in any association in this neighborhood? Have you expressed any opinions when you are living here? To who? For what?

Consumption Practices

13. Compared with your former neighborhood, how does the level of consumption change here?
14. What do you and your neighbours do in and around this neighborhood, for fun or for other activities? Where have you done these activities?
15. Where are your/your families' workplaces/schools now?
16. How do your family compare to others in this neighborhood—richer, poorer, the same?

17. Did you purchase the current apartment in the market or gain it through other ways, such as subsidised by the work unit? Can you tell me how much it costs and how did you pay for it?

Identity

18. How will you define yourself compared with others in the city, concerning cultural, social, economic and political capitals?
19. What do you think is the significance of one's living place and housing?
20. How will you compare yourself with others living in the luxury neighborhoods in suburbs?

For Retail Gentrifiers in Historical Neighborhoods Renovated for Commercial Use

Occupation

1. Can you tell me your educational background?
2. Do you have other jobs? Can you tell me the reason that you started your current business?
3. How does your family income compare to others in this city—richer, poorer, the same? Can you tell me generally your family's income band every year?

Residence and Locational Choice of Current Business

4. Where are living now? Do you rent or buy the apartment?
5. Do you have other housing properties?
6. Have you ever considered living in the current area, or anywhere in the central city? Why or why not?
7. Why did you choose to set up business in the current historic place?
8. Do you think the renewal of this historic place is successful?

Identity

9. How will you define yourself compared with others in the city, concerning cultural, social, economic and political capitals?
10. What do you think is the significance of one's living place and housing?
11. What dreams and goals do you have for yourself or your business?

For Residents Relocated by Money or In-Kind Compensation

Basic Information

1. Can you tell me your native place and education background?
2. Before living here, where did you live? Who were you living together with?

3. Why did you move here? Is the present apartment a resettled apartment due to the redevelopment of XX neighborhood?
4. How long have you been living in this apartment? What are the other members living in this apartment?

Previous Life and Working Experience

5. When was your previous neighborhood constructed? Do you know who constructed it?
6. How had it changed over years? How had the community members changed?
7. Which kind of jobs have you/your families done for a living? Can you tell me more about your previous jobs?

Project Participation

1. How did you first hear of the project to redevelop XX neighborhood? What were your initial feelings about this initiative?
2. Did residential representatives interpret for you the policies and regulations? For what kind of things?
3. Did you participate in any community meetings and events linked to the project? Why or why not?
4. Did you ever express your opinions on the project? If not, why? If yes, how have they been addressed?

Compensation/Relocation Information

5. Can you tell me how you have been compensated for the relocation?
6. Has all the compensation been delivered to you? Have you gotten any kind of commitments about the time and means of delivering the compensation?
7. Why did you/your families decide to remove and relocate with a compensation agreement?
8. Are you satisfied with the processes of project implementation and the compensation? Are there any worries in your mind?

Life in Existing Neighborhood

9. Do you and your families enjoy living in this apartment? Can you tell me more about your apartment and the decoration?
10. Do you interact with the neighbours here?
11. What's the difference compared with the previous neighborhood? Are you still in contact with neighbours in the previous neighborhood?

12. Have you participated in any association in this neighborhood? Have you expressed any opinions when you are living here? To who? For what?

Changes in Employment, Education and Consumption Practice

13. Where are your/your families' workplaces/schools now? How do you get there everyday?
14. Do they enjoy the current job or schooling? Have their jobs/schools changed over these years?
15. What do you do for fun in this neighborhood? What kinds of things do your family spend money on?
16. How does your family compare to others in this neighborhood—richer, poorer, the same? Can you tell me generally what is your family's income band every year?
17. Do you get an allowance?

Identity

18. How will you define yourself compared with others in the city?
19. Do you feel differently about yourself now from how you felt before or when you were younger? How?
20. What do you think is the significance is of one's living place and housing?
21. What dreams and goals do you have for yourself or your children?

For Private Tenants Who Had No Compensation and Were Displaced
Basic Information

1. How long have you/your families been in Chengdu?
2. Can you tell me your native place? Are you living in Chengdu by yourself or with your families?
3. When did you move to this apartment? Is the present apartment rented/purchased by you?

Working Experience in Chengdu

4. Can you tell me why did you come to Chengdu (studying, working or marriage)?
5. Which jobs have you done in Chengdu during these years?
6. Can you tell me more about your current job?
7. Did you get any kind of assistance from the government, such as the employment assistance?

8. How do you/your family spend your money? Do you save it? Do you send money to your families?

Residential Experience in Chengdu and Particularly in the Redeveloped Neighborhood

9. Before living here, where have you lived? How did you choose living locations?
10. Can you describe your previous apartments and neighborhoods, such as the living conditions, your neighbours, and rents in those places?
11. Why did your move out of the last neighborhood? Have you ever tried to stay put there? How?

Life in Existing Neighborhood

12. Can you tell me why did you choose to live in this neighborhood?
13. Do you interact with the neighbours?
14. Have you expressed any opinions when you are living here? For what? To who?
15. Do you have other housing properties in your hometown or Chengdu? Have you ever gained or applied for housing assistance?

Identity and Family Plans

16. What was the happiest moment of your life in recent these years?
17. Are your children at school in Chengdu?
18. What dreams and goals do you have for yourself or your children?
19. Have you ever planned to go back to your hometown? Why or why not?
20. How will you define yourself compared with others in the city? Can you tell me generally your/your family income band every year? Has that changed throughout the years you are in Chengdu?
21. What do you think is the significance of owning a private housing in Chengdu?

Ingram Content Group UK Ltd.
Milton Keynes UK
UKHW020638250523
422334UK00001B/1

9 789811 922886